ADDITIONAL PRAISE FOR *BEYOND THE FRONT LINES:*

"*Beyond the Front Lines* is a first-rate and much-needed explanation of the contemporary domestic, foreign, and global media in the crises-laden post-9/11 era. Illuminated by the rich insider knowledge of a former journalist and the healthy detachment of a journalism profes-sor, Seib's volume offers an authoritative account of the new challenges and responsibilities that journalists, news organizations, and govern-mental decision makers face in the changed geopolitical realities of the early twenty-first century. This well-written volume deserves not only the attention of journalists, students of communication, political sci-ence, and sociology but of the general public as well."

—Brigitte L. Nacos, Columbia University

"In *Beyond the Front Lines,* Philip Seib presents a highly readable and thought-provoking review of news media coverage of the 2003 war in Iraq. He examines journalistic triumphs and defeats and raises the tough questions that must be addressed by the news media before the next major U.S. military conflict. An outstanding work!"

—Kathy R. Fitzpatrick,
Attorney and Associate Professor and Director,
MA in Public Relations and Advertising, DePaul University

BEYOND THE FRONT LINES

How the News Media Cover a World Shaped by War

Philip Seib

First published 2004 by
PALGRAVE MACMILLAN™
175 Fifth Avenue, New York, N.Y. 10010 and
Houndmills, Basingstoke, Hampshire, England RG21 6XS.
Companies and representatives throughout the world.

PALGRAVE MACMILLAN is the global academic imprint of the Palgrave
Macmillan division of St. Martin's Press, LLC and of Palgrave Macmillan Ltd.
Macmillan® is a registered trademark in the United States, United Kingdom and
other countries. Palgrave is a registered trademark in the European Union and
other countries.

ISBN 1–4039–6547–1 hardcover

Library of Congress Cataloging-in-Publication Data
Seib, Philip M., 1949-
Beyond the front lines : how the news media cover a world shaped by war / by
Philip Seib.
    p. cm.
    Includes bibliographical references and index.
    ISBN 1–4039–6547–1
    1. War—Press coverage.   2. War—Press @Body_First:coverage—United States.
I. Title.

PN4784.W37S45   2004
070.4'333—dc22

                                                                    2003068929

A catalogue record for this book is available from the British Library.

Design by Letra Libre, Inc.

First edition: May 2004
10 9 8 7 6 5 4 3 2

Printed in the United States of America.

Dedicated to the memory of Charles Seib,
father and journalist.

*A great part of the information obtained in war is contradictory,
a still greater part is false,
and by far the greatest part is of a doubtful character.*

—Carl von Clausewitz, *On War*

CONTENTS

ACKNOWLEDGMENTS

The work of the women and men who covered the Iraq war and other conflicts is a journalistic genre that features much superb reportage. Added to this are the contributions of those who scrutinize the quality of wartime journalism—editors, critics, ombudsmen, and others—with an eye to making it even better. I benefited from the efforts of them all.

I also appreciate friends and colleagues who helped me more directly. At Marquette University, Bill Elliott and Ken Ksobiech were consistently supportive of my work. Some of the material in the book was presented in early form at conferences of the International Studies Association and my colleagues there offered helpful advice. Kathy Fitzpatrick and Viktoria Orlova read chapters in progress and provided valuable comments.

At Palgrave Macmillan, Toby Wahl has been a pleasure to work with. Editors with good ideas and flexible deadlines are to be treasured. My agent, Robbie Anna Hare, again proved her devotion to her writers and moved with speed and efficiency to help an idea become a book.

As always, my wife Christine advised and endured throughout the writing process. She sustained me.

PREFACE

This book was born during the 2003 Iraq war. As the war proceeded, I was surprised by the large number of news stories about how the conflict was being covered. For a profession that is usually reluctant to engage in timely introspection, this ongoing appraisal of the quality of wartime journalism was something of a breakthrough.

When the major fighting ended, many of the journalists who had been in Iraq wrote fascinating and often moving stories about their experiences. This genre will undoubtedly expand over time, but even the early appraisals of the journalists' work are useful reading for those who love or hate or just neutrally rely on the news media.

Reviewing all this early material got me started, and from there I looked at outlying issues: larger questions of press-military relations; the Internet as a news medium and as a battleground itself; the rise of Al-Jazeera and other international media organizations that are challenging the dominance of the West; and the efforts of the U.S. government to generate its own news product in public diplomacy programs designed to influence world opinion. Then I ranged still farther, considering the changing nature of war as well as war news, and examining the generally woeful state of the American news media's coverage of the world. These journalism issues affect policy making, so the evolving substance of American foreign policy also needed to be addressed.

When looking at the war coverage, I found much good journalism produced by courageous and thoughtful women and men, and I also found systemic flaws in how the news business works, particularly in terms of breadth and depth of coverage. These issues matter, because journalists, like policy makers, must always be planning for the next war and because the public depends on them to do their jobs well.

———

The Iraq war was part of a chain of high- and low-intensity conflicts that American forces will continue to fight and American journalists will continue to cover during the foreseeable future. As was seen in Iraq, the technologies of combat and news gathering continue to advance rapidly, and the relationship between those who fight and those who report also continues to evolve. For the most part, American troops and American journalists did their jobs bravely and well. Iraq was freed and the world was told about it.

Plenty of books are sure to be written about this war by the generals who led the charge across the desert and the reporters who forged a close bond with the troops. There is also a less dramatic but still important story that needs to be told that: despite the end of the Cold War, we find ourselves in a time of constant conflict that the news media are still deciding how to cover.

This story is more about ideas than battles:

- The post–Cold War era has given way to the post-9/11 era. Americans' feelings of triumph and security have been replaced by a fearful uncertainty about their safety that is accompanied by a vague recognition that the rest of the world cannot be ignored.

- America's machinery of war has reached the level at which victory can be achieved while sustaining relatively negligible losses. This fact may deter prospective enemies, but it also may encourage U.S. policy makers to rely more often on military options to advance American interests.

- New technology is redefining the mission and practices of warriors and journalists. "Smart" weaponry can wreak controlled havoc on an enemy. Its "smartness" should be accompanied by an ethical mandate to reduce civilian casualties and other incidental destruction that has been accepted as an unavoidable part of war. News gathering has also been dramatically enhanced by satellite communications and other tools, and the quality of coverage should rise to match the quality of journalists' high-tech gear.

- Globalization is changing the politics and journalism of conflict. A "clash of civilizations" may or may not be under way, but certainly new cultural and ideological rivalries have superseded those of the Cold War. For Americans, "the enemy" is much more amorphous than the Soviet bloc was, and part of the news media's job is to educate the public about the new causes and contexts of conflict as well as the dynamics of political and economic globalization. In the news business, globalization manifests itself in the expanded competition that Western media face. Al-Jazeera is a good example of this, and its impact on the Arab world during the Iraq war foreshadows a more volatile news environment that will profoundly affect foreign affairs as well as journalism.

- Use of the Internet is growing rapidly as people go on line to get information from news organizations and unmediated material from primary sources. Governments and terrorist groups are among those using the Web to disseminate information and enlist political support. The virtual landscape of the Internet also may become a battleground as cyberwar evolves from nuisance to significant threat, menacing the national and global electronic infrastructure.

These issues should be considered carefully because the next war might not proceed as neatly as did the push to Baghdad in Operation Iraqi Freedom. As events during the postwar war in Iraq have illustrated, creating and maintaining peace can be as challenging as major combat.

Another aspect of conflict today is terrorism and its soldiers may again bring devastation to American soil. Chemical, biological, and nuclear weapons are scattered around the world and they might be put to use. Opponents might not be as identifiable as Saddam Hussein was, and—as the first stages of the war against Al-Qaeda have shown—an enemy that is hard to find is hard to fight.

These challenges will confront political and military leaders and they must also be addressed by the news media. Fighting postmodern war is a complicated business and so is covering it.

American news organizations operate at a disadvantage because many of them have neglected international coverage in recent years. Reporters, editors, and producers do not know all they should about topics ranging from the global economy to global conflict, and the American news audience is largely ignorant about how the world works. This needs to be fixed.

The gadgetry of news is wonderful, allowing fast reporting from virtually anywhere, but most U.S. media organizations do not take full advantage of this. Much of their coverage consists of belated reports about crises, delivered with dramatic urgency but without context.

These flaws are strikingly visible in war coverage, which should not be seen as a stand-alone element of the news business, but rather as an integrated part of the ongoing effort to cover the world. Coverage of conflict should also reflect the changing politics of war—the messianic nature of current and potential foes; justifiable intervention versus respect for sovereignty; and the Bush doctrine that endorses preemptive self-defense. These issues were significant in setting the stage for the Iraq war and will continue to be important in the post-Iraq environment, but they have yet to receive the consistent attention from the news media that they deserve.

As part of the task of grappling with these big topics, some news organizations are trying to bring more consistency to the constituent elements of their coverage. Editors and producers have been working on the semantics of terrorism, trying to define actors and their acts accurately, without slipping into the jargon of propaganda. They are also addressing the use of visual images.

Can pictures of war be too graphic and therefore need to be withheld, or should the truth about combat never be sanitized? This gets to the essence of war's reality. For political reasons, governments like to portray war as a heroic exercise not far removed from Hollywood's version. Civilian casualties, friendly fire episodes, and mistakes of varying magnitude are not allowed to infringe on the dominant story line of grand achievement and noble sacrifice.

War, however, is much more than this, and anyone who thinks it is neat and clean is delusional. But some news executives are a lot like politicians, and as they read opinion polls and monitor audience attitudes, they slide along the path of least resistance, tidying up the mess of war and substituting flag-waving for journalism. This was an issue in the Iraq war, as it is in every conflict, and important questions remain about what version of reality—the government's or a more accurate, independent view—is delivered to the public.

During the Iraq war, the Bush administration's plan for the press demonstrated sophisticated understanding of journalistic practices and motivation. Recognizing the news media's unhappiness about the constraints on coverage that had existed during the 1991 Gulf war and realizing that the latest technologies would give roaming journalists unprecedented ability to report from wherever they wanted, the Defense Department decided that retaining some control over coverage was better than losing it all. The result was the Pentagon's embrace of the concept of the embedded journalist who would be assigned to a military unit for the duration of the war, live with the troops, and get a first-hand view of the fighting. News organizations were happy to get so close to combat and the Pentagon was happy about the way the embedded correspondents bonded with the troops they were accompanying. The product was vivid enough to grab the public's attention, but the perspective was so narrow that the audience was left with little understanding of the political dynamics that were behind the shooting and would shape the aftermath of the heaviest fighting.

Embedding grew out of a new approach to the relationship between the news media and the Pentagon. Ever since the Vietnam war, mistrust had permeated the marriage of necessity between press and military. At the heart of this was the myth that hostile news coverage was the reason the Vietnam war had been lost. Even the Pentagon's own studies of Vietnam coverage did not reach this conclusion, but the notion that "the media must be controlled" became an article of faith among many military officers and the civilians who command them. Restrictions on news gathering in Grenada, Panama, and the Gulf war exacerbated tensions between news organizations and the Department of Defense.

Rather than continue in this direction, Pentagon officials decided that "Improve journalists' access" would be their new mantra. In the spirit of détente, the Pentagon even sponsored survival training for journalists who would accompany the troops in Iraq.

Another influence on how a war is covered is patriotism. Close ties between American journalists and the American government reflect the idea that in wartime the adjective "American" takes precedence. That may be true, but should journalists be loyal primarily to the American government or the American public? Their interests sometimes diverge, but popular emotion during wartime can discourage telling politically unsettling truths. How the news media deal with this is a measure of journalistic fortitude.

Even during a brief war, the government will instinctively cover up or fuzz facts, sometimes for security reasons and often for political ones. Wartime requires special resolve on the part of journalists to not be pushed around, tempered by the recognition that operational security may occasionally require complicity with government secrecy. Those situations, however, are rare and journalists should keep the bar high.

While American news organizations were performing their minuet with the Pentagon, global news coverage was changing dramatically. Just as the Gulf war established CNN as an important player, the Iraq war enhanced Al-Jazeera's international standing as an alternative to Western news media. It became known as "the Arabic CNN," which is very different from "CNN in Arabic." The latter, regardless of the language being spoken, is still the product of an Atlanta-based news operation. Al-Jazeera provides news for Arabs by Arabs, with built-in cultural biases that are easy for outsiders to criticize but win the targeted audience's allegiance. This is *their* news.

Al-Jazeera is not unique. Shortly before the war began, Saudi-funded Al-Arabiya began its 24-hour news programming, joining other regional broadcasters whose ranks are certain to expand as technology costs decline and audience demand mounts. The Western news hegemony led by American media, the BBC, Reuters, and the like is ending, which is an important issue for governments as they court global public opinion. The din of many media voices—among them loud opposition voices—complicates politics.

Less dramatic than Al-Jazeera's rise to prominence but with greater long-term impact is the relentless growth of the Internet as disseminator of raw information and finished news products. By offering virtually infinite capacity and providing news on demand, the Web is winning adherents and chipping away at the dominance of traditional media. With the arrival of a new generation of news consumers that has grown up on line, news organizations are finding that convergence is becoming a necessity. If they are to capture this audience, they must offer an array of news products, on the Web as well as in print and on the air.

News organizations must also recognize that the Internet itself is newsworthy. Not just the news media, but governments and even terrorist organizations use the Web to proselytize. Also, the Internet may become a battleground in the next wars. Cyberwarfare is not science fiction; the first attacks have already occurred. So far, they have caused little harm, but in a society that is increasingly dependent on the Internet, hacking and larger-scale disruptions have the potential to be devastating. Much of the limited news coverage of this has so far been relegated to technology news niches. A wider public needs educating about this.

Public diplomacy is a more traditional kind of intellectual warfare. It received new attention from American policy makers following the September 11 attacks. After finding no ready answer to "Why do they hate us?" American officials realized that they had to do a far better job of presenting their case to "the Arab street" and other constituencies throughout the world. The news media have so far covered public diplomacy in an offhand way, with most coverage focusing on its stumbles rather than its purpose and the stakes involved. Criticizing public diplomacy as mere propaganda ignores the fact that the United States cannot afford to be forever cast as the world's villain. Allowing that to happen would mean allowing hatred to fester, with lasting antipathy and terrorism likely to be the result.

When considering these elements of present and future conflict, even the 1991 Gulf war seems ancient. Postmodern war is here, new in many respects but still ancient in its viciousness. The fighting and coverage of the Iraq war of 2003 gave us a look at some of what lies ahead.

CHAPTER ONE

Redefining the World:

Global Journalism and the Context of Conflict

When the United States began its war against Iraq in March 2003, the conflict seemed to many Americans to be inevitable and necessary. President George W. Bush and members of his administration had skillfully made their case for war and pulled public opinion along with them. Much of the rest of the world was dismayed, but such feelings barely registered, in a domestic political sense, among Americans. Debate and protests about the war within the United States developed only limited emotional momentum, and they failed to capture widespread attention. Concerns about the long-term effects of the war on America's role in the world spurred little discussion. As opinion polls showed, Americans had decided that they wanted this war and didn't want to be bothered about details.

But how did they arrive at that decision? On a matter as important as going to war, thorough debate is essential, and the news media should be among the instigators of such debate. News organizations should encourage questions about the official version of events and should provide an ample supply of information that news consumers can use to make up their minds about what they think their country should do. If the news media do not provide those tools, an imbalance exists that undercuts democracy. What should be a healthy dynamic tension between political leaders and the public atrophies, and the public sits back while the government does whatever it wants.

If the news media are consistently doing their job, the public will have an ever-growing reservoir of fact and opinion that they can use in making their decisions about the need for war, as well as about other policy matters. In 2003, that was not the case. The U.S. news media's prolonged weak performance in covering the world left the public underinformed, and America went to war amid circumstances that merited closer scrutiny.

That is all in the past. Now news organizations must decide how to cover wars yet to come—the events leading up to them and their aftermath as well as stories from the battlefield. Much needs to be done to lift this coverage to the level of excellence that the public deserves.

━━━━━━━━

Surveying international news coverage by the American media can be dispiriting. Much of the news product has a narrow perspective, as if the world today was the same as that depicted on maps from the Middle Ages: a handful of identified territories, much terra incognita, some roving sea monsters, and then a sheer drop off the edge of the flat earth.

In the coverage that *is* available, parochialism flourishes, as if the politics and culture of a news organization's home country were all that matter. The consumers of this product, not surprisingly, tend to adopt a similarly insular outlook. They are curious about little and care about few.

This is a familiar but still important indictment. The flawed journalism is sustained by a self-perpetuating cycle. News organizations, short on initiative and money, cut back their international coverage. News consumers don't miss what they don't know about. Seeing little coverage of the rest of the world, they assume that there is nothing worth reporting and they remain content with their diet of close-to-home scandal and the foibles of the celebrities of the moment. Audience surveys commissioned by media organizations then document this lack of interest in the world and the findings are used to justify further cuts in international reporting. Coverage shrinks, demand shrinks, and the quality of news judgment and reporting diminishes as well. This branch of journalism withers as does the intellectual ambition of news consumers. We are left with stories such as Winona Ryder's shoplifting trial as the centerpieces of news.

A good way to begin an examination of the news media's performance is to ask a few questions:

- Are news organizations looking beyond the government's priorities and covering matters that policy makers neglect?
- Does news coverage provide the public with the information needed to evaluate government's performance in a timely way or does it lag behind events and scrutinize policy only after decisions have been made and full-blown crises are under way?

- Is the full range of technology-enhanced news venues—from print to Web—being used to provide the public a broad spectrum of information?

The answers to these and related questions are grounded in fundamental principles of journalism and reflect recent developments in the science of delivering the news.

Scope of Coverage

The gadgetry is wonderful. In addition to their seemingly magical powers to help gather, assemble, and disseminate news, the latest advances in communications technologies can expand the scope of news coverage. The tools include ever-smaller and more mobile satellite uplinks; satellite telephones, including videophones; and satellite photography.

These improvements in gathering and delivering news allow journalists to report from places where timely coverage used to be almost impossible. This should mean that there will be fewer invisible wars and fewer forgotten peoples. In 1938, Neville Chamberlain dismissed Czechoslovakia as "a far away country . . . people of whom we know nothing." Today, if the news media are doing their job, "knowing nothing" about another country's people will not be a viable excuse for a government's abandonment of that nation. No country and no people will be beyond journalistic reach. Citizens of nations that most people cannot even find on a map will suddenly, vividly appear in living rooms throughout the world.

The breadth of news coverage depends on news organizations' view of the world, a view that is often too narrow. Expanding it will require a surge of ambition and a reversal of the reductions in international coverage. Media analyst Andrew Tyndall reports that in 1989 the ABC, CBS, and NBC evening newscasts presented 4,032 minutes of datelined coverage from other countries. That had dropped to as low as 1,382 minutes in 2000. With the attacks on the United States and the war in Afghanistan, the figure rose to 2,103 minutes in 2002, which is still only slightly more than half the total of 1989.[1] Air time devoted to international stories not involving U.S. policy dropped from 4,828 minutes in 1989 to 2,937 minutes in 2002.[2]

One aspect of the shrinkage of international coverage is the reduction in the number of foreign bureaus maintained by American news organizations, notably the big three television networks. As of mid-2003, ABC, CBS, and NBC each maintained six overseas bureaus with full-time correspondents, but since the peak of international coverage during the 1980s, each has closed bureaus or removed correspondents when there was not a full bureau in place. ABC did this in seven cities, including Moscow, Cairo, and Tokyo. CBS did it in four cities, including Beijing and Bonn. NBC followed suit in seven cities, including Paris and Rome.[3]

The weakness of international coverage is no secret within the news business. A 2002 study conducted for the Pew International Journalism Program found that among American newspaper editors, "nearly two-thirds of those responsible for assembling their newspaper's foreign news coverage rate the media's performance in this area as fair or poor."[4] When asked about their own news organization's performance in satisfying readers' interest in international news, 56 percent gave their own paper a rating of fair or poor (and only 2 percent rated their paper as excellent).[5]

Editors at newspapers with a circulation of at least 100,000 were particularly critical of television news. Sixty-seven percent of the editors said network television news did a fair or poor job of covering international events, while 40 percent said cable news coverage deserved only a fair or poor rating.[6] Overall, the study found, "the ratings given to international news coverage were significantly lower than those awarded to the media's coverage of sports, national, local, and business news."[7]

Such lackluster performance stands in contrast with what the editors perceived as an increase in the public's interest in international news, which contradicts the conventional wisdom that the news audience resists learning about the rest of the world. In general, said the editors, only 7 percent of their readers are not too interested in international news.[8] Ninety-five percent of the editors said reader interest in international news had increased since the September 11, 2001 attacks, but 64 percent said that they believed that this interest would soon decline to pre-September 11 levels.[9] This reflects a condescension on the part of journalists toward the public that in itself merits study, particularly in terms of the values governing the relationship between news media and the people they purportedly serve.

Another survey, conducted for the Project for Excellence in Journalism, found that by spring 2002, network television news had largely reverted to its pre-September 11 line-up of topics. The amount of hard news shrank, from 80 percent of stories in October 2001 to 52 percent in early 2002. Meanwhile, the number of "lifestyle" stories made a comeback. Such stories made up 18 percent of total network news stories in June 2001, 1 percent in October 2001, and back to 19 percent during the first 13 weeks of 2002.[10] This continued a trend that has been noticeable for more than a decade.

These findings indicate that in this age of globalization, when the news media's view of the world could and should become ever broader, intellectual isolationism has taken hold, at least in journalism and presumably other fields as well. When asked what obstacles kept them from increasing international coverage, 53 percent of the editors in the Pew survey cited cost. This was followed by lack of interest by senior editors and lack of experienced reporters, each cited by 9 percent of the editors.[11]

If improvements are to be made in international coverage, the process should include looking anew at the world's political and geographical structures, which will entail reevaluating traditional boundaries and alignments. Even semantic changes may be significant. Philosopher Peter Singer said that "implicit in the term 'globalization' rather than the older 'internationalization' is the idea that we are moving beyond the era of growing ties between nations and beginning to contemplate something beyond the existing conception of the nation-state."[12]

Some of this reappraisal should be based on new economic realities. Business executive Kenichi Ohmae said that "the nation state has become an unnatural, even dysfunctional, unit for organizing human activity and managing economic endeavor in a borderless world." He makes the case for the rising dominance of "region states," which connect directly to the global economy.[13]

Other descriptions of a revised global structure range beyond conventional political and economic topics. In *The Clash of Civilizations and the Remaking of World Order*, Samuel Huntington argued that "the most important distinctions among peoples are not ideological, political, or economic. They are cultural."[14] Huntington's thesis has spurred much debate, which in itself has been useful in stimulating reconsideration of what lines—physical or otherwise—divide the world. He said that "states are and will remain the dominant entities in world affairs,"[15] but he also noted that intraregional relationships, grounded partly in broad cultural commonality, are of growing importance. He cited the European Union and NAFTA as examples.

It is easy to take some of the ideas in Huntington's book, such as his perceptions of the Muslim world, and stir up an argument about the significance of "cultural fault lines" and other factors that may foster increasing contentiousness throughout the world. International relations scholar Charles Kupchan said that "there is scant evidence to support the claim that other cultures will clash with the West" and notes the absence of resentment-fueled "coalitions among non-Western civilizations."[16] Of course, some members of those civilizations—even if not the civilizations collectively—may well think themselves to be representative of the larger entity and instigate clashes with the West. In individual cases, questions will arise about the degree to which these aggressive groups reflect sentiments of the whole. Does Al-Qaeda, for example, represent a large constituency, or is it merely a small, if sophisticated, criminal enterprise that cloaks itself in religious and cultural rectitude?

Religion's potency is an integral factor in global tensions and may supersede traditional political interests. Middle East historian Bernard Lewis said it is important to understand that religion can be "subdivided into nations rather than a nation subdivided into religions—and this has induced some of us to think of ourselves and of our relations with others in ways that had become unfamiliar. The confrontation with a force that defines itself as Islam has given a new relevance—indeed, urgency—to the theme of the 'clash of civilizations.'"[17] The events of September 11, 2001, and the conflicts that followed underscored the importance of not only dealing with states such as Iraq, but also scanning the horizon to take note of non-state entities.

Regardless of what forces are clashing, debate about such matters helps nudge thinking away from acceptance of the status quo. That is an essential step not just for the public and scholars, but also for journalists as they reappraise their coverage responsibilities.

Taking a new look at the world should include a reconsideration of the role of maps. Enhanced by satellite photography and other technological aids, the modern map is considered an exceptionally reliable representation of the world's constituent parts. But that may be too facile a judgment. Maps lack political permanence. Scholar Benedict Anderson said that maps may be mere reflections of "the late colonial state's style of thinking about its domain." The map, he said, was part of a "totalizing classificatory grid, which could be applied with endless flexibility to anything under the state's real or contemplated control: peoples, regions, religions, languages, products, monuments, and so forth."[18] In this regard, the map is a relic of a proprietary assertiveness that is likely to prove increasingly distorting as years pass. Policy makers and journalists should be wary about putting too much faith in maps as definitive representations of geopolitical reality.

Today's map of the world, said journalist Robert Kaplan, "is generally an invention of modernism, specifically of European colonialism. . . . The map, based on scientific techniques of measurement, offered a way to classify new national organisms, making a jigsaw puzzle of neat pieces without transition zones between them. 'Frontier' is itself a modern concept that didn't exist in the feudal mind. And as European nations carved out far-flung domains at the same time that print technology was making the reproduction of maps cheaper, cartography came into its own as a way of creating facts by ordering the way we look at the world."[19]

These "created facts" may reflect a politically appealing version of reality that differs from "real reality." Most maps of the world show formally recognized nations. There is, for example, no Kurdistan on the most widely used maps. But an argument can be made that there *is* a de facto Kurdistan that is as real as any graphically depicted state and that should be recognized as a state-like subject for news coverage. It is a cultural entity of roughly 20 million people that spreads over parts of Turkey, Iraq, Iran, Syria, and former Soviet republics.[20] The existence of a Kurdish political identity has been implicitly recognized by the United States and others since the aftermath of the 1991 Gulf War, when U.S. forces created a protected zone for Kurds in northern Iraq. This was also an issue during and after the 2003 U.S. invasion of Iraq, when many Kurds made the argument that a considerable chunk of territory should belong to them. Other factors, such as the role of Turkey, were involved, but this complex case illustrates how the cartographic status quo may differ from political reality.

Even without official recognition, Kurdistan can be found. Kurds have adopted their own maps, which they believe should replace those in common use, and they have defined their own borders, which they contend should replace boundaries that are so antiquated as to be meaningless. Kaplan wrote about visiting the Turkish-Iranian border and finding Kurds on both sides of a political line that can be pointed to on a map, but that has little practical significance beyond turning conventional commerce into smuggling.

Kaplan cited the importance of "the political and cartographic implications of postmodernism." He said that this will be "an epoch of themeless juxtapositions, in which the classificatory grid of nation-states is going to be replaced by a jagged-glass pattern of city-states, shanty-states, [and] nebulous and anarchic regionalisms."[21] Instead of borders, he said, "there would be moving 'centers' of power, as in the Middle Ages. These power centers would be both national and financial, reflecting the sovereignty of global corporations. . . . Henceforward the map of the world will never be static. . . . Because this map will always be changing, it may be updated, like weather reports."[22]

Kaplan's ideas present a challenge to policy makers and others whose work is grounded in the "grid of nation-states." This includes many journalists, whose reporting assignments are frequently the offspring of a marriage between maps and airline schedules: the definitions of the map are readily respected as are the bounds of rapid accessibility that the conventional travel system dictates. If news coverage is to move beyond this, it must be planned in ways that recognize geography's revised realities.

Covering the Muslim world brings some of these realities into focus. Middle East scholar John Esposito wrote that "many of us have forgotten what the twentieth-century map of the Muslim world reveals. The names of regions (the Middle East) and countries as well as the boundaries and rulers of countries were created by European colonial powers." He added that "nation building in the Muslim world with its artificially drawn borders superficially uniting peoples with diverse centuries-old identities and allegiances was a fragile process that bore the seeds for later crises of identity, legitimacy, power, and authority. When we ask today why much of the Muslim world remains politically unstable or underdeveloped, we need to remember that most modern Muslim states are only several decades old, carved out by the now-departed European powers."[23]

This leads to an issue concerning news coverage of the Muslim world: whether coverage should focus on states or on the more amorphous *ummah*, the worldwide Islamic community. This latter approach presents special difficulties. Given the contentiousness that flares among Islamic factions, treating that community as a cohesive whole can be misleading. Esposito said that "like tribal or ethnic communities and nation states, they often pull together when faced by a common external threat but then fall back into intrareligious conflict."[24] So when the news media treat "Islam" as a cohesive religious-political entity, they risk slipping into oversimplification and stereotyping, which the public then adopts and perpetuates.

Despite the challenges inherent in designing coverage of the Islamic community, it deserves higher priority from news organizations partly because the unifying power of information technology is doing much to reinvigorate the concept of the *ummah* as a supranational entity. Easily accessible information, such as the news presented by CNN, the BBC, and more recently Al-Jazeera and other television organizations with global reach, can foster solidarity within an audience. The Internet carries this even farther, providing not just a flow of information but also an interactive medium that can bring unprecedented cohesion to the most far-flung community. Scholar Gary R. Bunt has noted that "it is through a digital interface that an increasing number of people will view their religion and their place in the Muslim worlds, affiliated to wider communities in which 'the West' becomes, at least in cyberspace, increasingly redundant."[25] As the Internet continues to reduce the significance of national borders and other boundaries, the entire array of global media and information technology may help create virtual communities that are as worthy of coverage as traditional states have been.

These new political realities and advances in communication technologies necessitate reassessment of the scope of news coverage. To provide accurate, comprehensive journalism, news organizations must address structural reconfigurations and social and political realignments among states and peoples. The news media should take the initiative in this, rather than waiting for policy makers to alter *their* approach to the world, which too often is merely reactive, consisting of unplanned responses to unanticipated events.

Questions about the validity of borders are nothing new. Among the antecedents of today's border-related disputes was the work of those who created boundaries after World War I. As he worked in Paris in 1919, British diplomat Harold Nicolson observed, "How fallible one feels here," but he proceeded to draw borders with his pencil and tracing paper. He added that "my courage fails at the thought of the people whom our errant lines enclose or exclude."[26]

The resulting difficulties are particularly acute where the imprint of colonial powers remains most visible. Boundaries in central Africa, for instance, were drawn for the convenience of European interests and resulted in some "national" populations that feature more internal antipathy than commonality. Between 1880 and 1900, 10,000 African tribal kingdoms were merged into 40 states, 36 of which were under direct European control.[27] The bloody misery that resulted from this has not only scarred history, but has continued to perplex the news media. Coverage of this area, even when timely, has tended to be superficial and even misleading. Western media, in particular, are so ready to rely on frames of reference that are appropriate to their own cultures that they fail to understand the very different realities of the peoples and situations they are attempting to cover. As a result, horrific conflicts in Sudan, Rwanda, Congo, Liberia, and elsewhere have been treated as mysterious phenomena and have not been reported in ways likely to capture the interest of the public or policy makers in countries that have the muscle, although rarely the will, to intervene and save lives.

The journalism that does emerge may be simplistic, condescending, and sometimes have overtones of racism. News coverage can easily slip into stereotyping and a public that is not particularly knowledgeable about the world may accept such images without questioning them.

Guarding against that intellectual sloppiness is an important task for news organizations. Such watchfulness requires increased sophistication in all aspects of international news gathering—planning the coverage, reporting, editing, and packaging the news product. News stories should reflect the complex reality of their subjects, because to provide anything less encourages the public to quickly dismiss the events being covered as unimportant.

Creating a more up-to-date approach to coverage should include reevaluating even the most traditional concepts, such as the relevance of sovereignty.

Legal historian Philip Bobbitt suggested five developments that call into question the conventional model of the sovereign state:

- The recognition of human rights as norms that require domestic adherence by all states, regardless of their internal laws.
- The development of nuclear weapons and other weapons of mass destruction that render the defense of state borders ineffectual for the protection of a society.
- The emerging recognition of other global and transnational threats that transcend state borders, such as those that damage the environment, or threaten states through migration, population expansion, disease, or famine.
- The growth of a world economic regime that ignores borders in the movement of capital investment to a degree that effectively curtails states in the management of their economic affairs.
- The threat to national cultures posed by the revolution in international communication, linking all cultures to one language that competes with local forms and penetrates borders electronically.[28]

These are complex political topics. It may not be journalists' role to pass judgment on sweeping political theory, but it is part of the job of news organizations to be aware of thinking in academic and other circles that offers new perspectives and insights about the world that the news media cover.

On a more pragmatic level, the logistics of coverage should never become static. Bureau locations and assignments can be shifted and redesigned to reflect changing reality. The scope of coverage should include virtual states such as Kurdistan and virtual communities such as the *ummah*. Traditional maps should not be discarded, but nor should they be accepted as defining the limits of political reality. Throughout the world, journalists should view their subjects of coverage with a vision that is not rendered myopic by convention.

Looking at the world differently will produce a different-looking journalism. That will be a step toward educating the world about new realities.

Speed of Coverage

Many of today's news organizations are determined to deliver not just news, but *real-time* news from anywhere and everywhere. This reflects the technology-driven evolution of the news business that can be seen in the coverage of recent wars. The Vietnam war was the first conflict to be televised day after day; the 1991 Gulf war was covered live; the Kosovo war was the first in which the Internet was an important tool; and the 2001–2002 fighting in Afghanistan was covered through the use of mobile, satellite-enhanced news gathering. During the 2003 Iraq war, television, radio, and Web journalists were able to deliver their reports even faster.

The conflict in Afghanistan provided a testing ground for a new generation of technology. *The Lion's Grave,* a compilation of reporting from Afghanistan by *New Yorker* correspondent Jon Lee Anderson, includes e-mails between Anderson and his editor that not only underscore the rigors of combat reporting, but also illustrate the level of reliance on satellite phones and such. In one e-mail, for example, Anderson discussed the virtues of a laptop computer with "extra-longlife battery and superduper casing (bulletproof)." He also reported that he could obtain an "Inmarsat [satellite] phone that should solve our communication problems. 2,500 bucks."[29]

Real time is the increasingly common standard throughout the news business, not just for TV and radio. As print media upgrade their on-line product, they too want real-time capability. This produces an array of challenges to news systems and standards. When stressing speed, news organizations may be tempted to cut corners in what should be a deliberative process of judging newsworthiness and determining accuracy. Newsroom staff members scan television and computer monitors, watching the competition, and reacting immediately. Even if their version of the story is not quite ready, they may rush it onto the air or onto their Web site in order to keep pace with the competition.

When the world is watching, effects of real-time coverage increase and so does responsibility. Inaccuracies don't simply vanish into space; they have impact. Particularly during crises, people rely on news reports and expect an accurate picture of what is going on. During the Iraq war, the first, fast reports did not always hold up. Infiltrators attacked a U.S. camp in Kuwait; no, it was a U.S. soldier. Iraqi divisions were surrendering en masse; no, just some smaller units were. Basra was secure; no it wasn't; well, maybe parts of it were. Iraqi crowds were applauding American troops; no, they were shaking their fists; no, they were doing both.

Does it matter that a news story is wrong? In these instances, the world did not tremble, no one's safety was jeopardized, and better information was soon provided. But the public had been misinformed, and that is no small thing. Sloppy reporting undermines the trust that news consumers should be able to place in journalism. Every error eats away at believability, like termites chewing through wood. The damage may be hard to detect initially, but the structure— the relationship between news media and public—is weakened. The vast majority of reporting from Iraq was accurate, but the exceptions were harmful.

Almost all of the errors could have been avoided if a bit more time had been taken to check out the information before delivering it.

Newspapers have joined television and radio news organizations as players in the real-time game, willing to scoop their own on-paper product by posting stories early on their Web sites. One newspaper editor said, "We're now on the same schedule as CNN." That is an interesting observation, but it is not clear whether it merits applause or worry.

Because delivery of the news product has become constant rather than periodic, the newsroom's rhythm has changed, as have its procedures. Consider the routine of the newspaper newsroom and how it has been altered: from a measured pace of gathering, analyzing, verifying, expanding, verifying some more, and refining a daily product; to instead delivering whatever is available at a minute-by-minute pace. Certainly, newspaper newsrooms have always had to deal with stories that break just before (or after) deadline, but the new routine is a driving force that makes its presence felt all day every day. Reporters are expected to have material ready for the Web and, at a growing number of newspapers, for broadcast from a permanent television camera position in the newsroom.

Those who work in this accelerated news process may find it tempting to set aside judgment in return for speed. To avoid that, reporters and editors must master more background and contextual information that will help them reduce chances of making bad decisions about when to go with a story. Just as the pace of news has quickened, reaction to news reports also happens faster. Every major government monitors not only CNN and its TV brethren, but also the Web sites of *The New York Times,* the BBC, and others. For policy makers, time for analysis and decision making is compressed. They may resist being stampeded, but it is hard to ignore news coverage and its potential impact on public opinion.

Consider, in contrast to today's pacing, a case from long ago. Berlin, the early hours of Sunday, August 13, 1961. CBS News correspondent Daniel Schorr and producer Av Westin are summoned from a nightclub by one of their sources; something unusual is going on near the Brandenburg Gate. When the CBS crew arrives, they see a large deployment of East German soldiers, some of whom are using jackhammers to dig holes for concrete posts. Barbed wire and barricades are being put into place. This is the birth of what was to become the Berlin Wall.

Nearby, American, British, and French troops are standing by in full battle gear; U.S. tanks and armored cars have moved up and wait with engines idling. Spotlights are playing on all this. The dramatic scene is perfect for TV.

But it does not instantly appear on the air. Schorr's crew films what is happening, and later in the morning the film is taken to Berlin's Templehof Airport for a prop flight to London. The plane gets there on time, so the film makes the London–New York jet, which arrives at New York's Idlewild Airport late Sunday afternoon, New York time. A courier picks up the film and takes it to a processing lab. Then it finally is delivered to the CBS newsroom. As producer Av Westin remembers it, the footage did not air until Tuesday night.[30]

Meanwhile, word of events in Berlin moved slowly within government circles. Twelve hours passed before a report reached President John F. Kennedy, who was at his home in Hyannisport, Massachusetts. After receiving a telex from the State Department, Kennedy consulted with Secretary of State Dean Rusk by phone, issued a bland statement, and then went sailing. He was purposely keeping the temperature low as he waited to see how events unfolded.[31]

By early Monday, news reports had informed the public about this latest Berlin crisis, but the dramatic television footage had not yet aired. The public and the press corps remained calm. At the day's regular White House news briefing, 34 questions were asked before Berlin came up.[32]

Consider how that situation would play out with today's technology. Live coverage from Berlin on CNN and other networks would pour into American living rooms during Saturday night prime time. With those images and accompanying speculation and commentary, the "crisis" would be much more pronounced. "Confrontation in Berlin" logos would pop up on TV screens; network anchors would don their flak jackets. The president would face sharp press queries and expectations that he would "do something" at once.

Kennedy was able to move at his own pace partly because he faced no immediate heat generated by news coverage. He and his advisors could watch and wait as they planned America's response. Today, with news reports pushing the pace of decision making, presidents and other policy makers must be prepared for the pressure that this real-time coverage can produce.

These issues may be collectively grouped within the rubric of the "CNN effect," which is generally defined as news coverage—real-time and other—driving policy. The concept, however, is often treated too casually, particularly by some scholars who overstate the impact of coverage. At the heart of this problem is the failure to distinguish between "influential" and "determinative." Certainly, news coverage can be one among many factors that *influence* policy makers, but to argue that coverage in itself actually *determines* policy reflects an overestimation of media power. The coverage is just one of many factors that policy makers take into account as they do their jobs.[33]

That is not to say that government officials disregard the impact of news. Speaking about the American intervention in Somalia in 1992, Secretary of State Lawrence Eagleburger said that "television had a great deal to do with President Bush's decision to go in the first place . . . it was very much because of the television pictures of these starving kids." Eagleburger also cited pressures from members of Congress and administration officials' belief that the mission would be low-risk in terms of U.S. troops incurring casualties.[34]

Television news can also be a useful tool in diplomatic communication, to a certain extent replacing the diplomatic pouch. During the 1991 Soviet coup, when President George H. W. Bush decided that official U.S. policy would be to support Mikhail Gorbachev's restoration and Boris Yeltsin's efforts to rally public opposition to the coup, his medium for conveying his views was a televised news conference. Secretary of State James Baker said that Bush "had used the fastest source available for getting a message to Moscow—CNN."[35]

Real-time media also may lend itself to disinformation—the purposeful dissemination, for policy-related purposes, of information known to be false. Government officials are adept at dispensing tantalizing slices of information to reporters. A leak from a high-ranking military official may be a bluff, but with the imprimatur of a major news organization it will at least command attention. As American forces were tightening their hold on Iraq in April 2003, news stories noted the possibility of U.S. military action against Syria. Such a move was not actually in the works, but the reports did what the Bush administration wanted them to do: help nudge the Syrians into being more cooperative. This kind of gamesmanship is not new, but news organizations are more susceptible to such manipulation when desire for speed outweighs concern about verification. Journalists should recognize that the job of the news media is not to simply be a conveyor belt, delivering whatever is dumped onto it. In a real-time news environment, however, suspect items might not be pulled off the belt and the crucial process of verification and analysis of news (or what purports to be news) may be shortchanged.

In many ways, communications technology is bringing those who deliver the news and those who make policy closer together. Because news organizations get information so quickly and have access to tools such as satellite photography, governments should understand that their ability to affect the flow of information and the context of news is less one-sided than it once was.

This is a kind of convergence—not that which is usually discussed as being a technological merging of media, but rather a narrowing of the gap between professions. Despite the dynamic tension that rightly shapes much of their relationship, journalists and policy makers need to better understand how the practices and standards of each affect the other.

For news organizations, this includes devising reforms similar to those initiated after the bumbling coverage in 2000 of the U.S. presidential election returns. Reducing speed and increasing precision might improve coverage of foreign affairs as well as electoral politics.

The news media's commitment to provide the best obtainable version of the truth does not require so much caution that the news will be whittled down to only the least debatable snippets of information. Rather, it imposes a standard that news organizations can meet if speed is not in itself driving their coverage. That is a simple but important test.

Diversity of News Sources

When considering these issues, it is important to keep in mind that the global news universe is constantly expanding. Looking at the status of Internet use today, we may be at a technological watershed comparable to the mid- or late-1950s in the development of television. Although many of the early Internet-related expectations—especially economic ones—have not yet been met, their time will come. For one thing, a tidal wave of Internet users in the next generation will soon hit. Children today are using the Internet as regularly and casually as their parents used television and the telephone when they were growing up.

Change also continues among traditional American media. The big three television networks' newscasts lost 40 percent of their viewers between 1981 and 2001. CNN, which established itself as a major news organization during the 1991 Gulf war, has been overtaken in the ratings contest by "edgier" Fox News. Newspapers and news magazines watch their circulation continue to shrink. The economic bite that accompanies these changes has had a pronounced effect on international news coverage, as overseas news gathering has been reduced to save money.

Another significant change has been the challenge to Western news media dominance exemplified by the rise of Qatar–based Al-Jazeera. Operating since 1996, with financial support from Qatar's Sheik Hamid bin Khalifa Al-Thani, this satellite news service covers the world from an Arab perspective. Its staff includes more than 350 journalists, with foreign correspondents in 31 countries, and its viewership has been estimated at more than 35 million and climbing.[36]

During the Iraq war, Al-Jazeera aired graphic video of civilian casualties and thoroughly reported anti-war protests in the Arab world and elsewhere. This coverage, and that by other Arab media, appears to have substantial impact on Arab public opinion, which has made American public diplomacy—the delivery of the U.S. message about its policies—more important and more difficult. (A closer examination of Al-Jazeera's impact, particularly during the Iraq war, will be found in chapter six.)

Other emerging media voices are finding that the Internet, rather than television, provides a good home. IslamOnline, for example, offers coverage that is clearly tilted in favor of Palestinian and Muslim interests, and is critical of Israel and American Middle East policy. It appeals to a broad audience, supplementing its news coverage with discussion pages, a "Fatwa Live" feature that allows site visitors to ask for religious judgments about personal and wider issues, and even a matrimonial service. The material is presented in Arabic and English because only a quarter of the world's 1.2 billion Muslims speak Arabic, while many know English. It has editorial offices in Egypt and corporate headquarters in Qatar. Like Al-Jazeera, it receives financial support from Qatar's royal family.[37]

The expanding number of news organizations makes possible new media relationships. A partnership between the BBC and Al-Jazeera involves logistical and technical cooperation, such as the BBC being able to use Al-Jazeera's facilities in Kabul, Afghanistan. Yet to be determined is whether these relationships will go farther and include some integration of their coverage. If this happens, perhaps parochialism will be diminished and news audiences will benefit from exposure to the different viewpoints of the partnering organizations.

On a wider scale, news consumers are increasingly taking advantage of unmediated media. People with Internet access can read the world's newspapers, watch the world's television, access the world's governments, and hear from groups and individuals previously lacking means to amplify their voices. They can go themselves to primary sources, rather than relying on news organizations to gather and filter information. Such sources range from government bodies such as North Korea's official news agency and America's White House, to NGOs such as Doctors Without Borders, and to terrorist groups such as Hezbollah.

As more of the world goes online, news-related uses of the Internet will expand. For instance, the non-partisan Center for Defense Information, which is based in Washington, D.C., offers Washington ProFile, a free, on-line digest of American news translated into Russian three times a week and into Chinese once a week. The material is accessed by news organizations and individuals, and the staff hopes to expand the service into other languages, such as Farsi, Arabic, and Spanish.[38]

Public access to the vast and growing reservoir of on-line material challenges the role of traditional news media. Despite the appeal of getting information on their own, news consumers should note the important difference between "information" and "journalism." Journalism is, in part, the application of judgment to information—sifting through the pile, verifying, analyzing, explaining, going the next step. Most news consumers do not have the time, inclination, or expertise to do all that on a continuing basis. That can leave them open to manipulation by governments and others that are more adept at packaging what they label as "news" than at telling the truth.

For nations, organizations, and individuals, the Web is an alluring propaganda tool. DefendAmerica.mil, for example, is used by the Department of Defense to deliver material about the war on terrorism in the form it chooses. Whether such sites are informative or manipulative depends partly on the pre-existing opinions of the site's visitors. Will the public trust and rely on this kind of government-produced information rather than on what the news media provide? If so, how will that affect political support for various government policies? How will the news media respond in order to keep their audience?

Answers to these and related questions are just starting to take shape.

Developing a New Approach to International Coverage

International news coverage can be improved and journalists have the ability and responsibility to do so. This is not merely a matter of news business mechanics—opening new bureaus, spending more money, and expanding the air time and column inches allotted to international news. It also requires a recognition of changes in the world's politics and an understanding of how such changes reshape journalism's mission.

Too much of the system of international news coverage remains mired in the archaic conventions of the bipolar Cold War world, when simplistic foreign policy was often matched by simplistic news coverage. The world's players—at least those judged to be worth reporting about—were divided into good guys and bad guys. That was the context in which decisions were made about which stories were important.

After almost a half-century of seeing the world conveniently divided in that way, the news media today confront a more amorphous international community. Even today's "bad guys" (as defined by Western media), such as Al-Qaeda, may have no home that can be identified on a map. New communities of interest, such as the European Union and Mercosur, make coverage of transnational entities important. Globalization takes that a step farther, as supranational economic and political interests become more significant. Giant corporations transcend nationality and are governed through cyberspace. Humanitarian emergencies in remote places that would have escaped notice in the past now come into the world's living rooms as "virtual" crises.

Such changes challenge policy makers and journalists alike. While governments decide how to adapt to these new realities, the news business must realign its own priorities if journalists are to help the public develop a better sense of what is going on in the world.

———

War, like the news media that cover it, is changing in its technology, purpose, and effect. For the news media, the Iraq war was a transforming event, much as the Vietnam war had been during the 1960s and '70s. But before examining the Iraq war, it is useful to consider war and war coverage as they evolved during the period shortly before the conflict of 2003.

CHAPTER TWO

Prelude to Iraq:

The Changing Nature of War and War Coverage

Forget the macho tales of spit-shined glory, war is death and misery. Despite "revolutions in military affairs," shifts in the dispersal of power throughout the world, and quantum leaps in technology, the fundamental nature of war has not changed. The progress of humankind has been remarkable in many ways, but an intrinsic viciousness remains and surfaces persistently in conflicts throughout the world.

What *has* changed is the scope of devastation and the nature of war's victims. At the beginning of the twentieth century, the ratio of military to civilian casualties was 8:1. By the 1990s, it had become 1:8. Scholar Mary Kaldor says that "behavior that was proscribed according to the classical rules of warfare and codified in the laws of war in the late nineteenth century, such as atrocities against noncombatants, sieges, destruction of historic monuments, etc., now constitutes an essential component of the strategies of the new mode of warfare."[1]

Perhaps the newest generations of weaponry, the "smart" bombs and arms that use pinpoint targeting, will start making a difference. Military targets can be so precisely pulverized by aircraft and missiles that broader attacks may not be necessary. During the first stages of the 1991 Gulf war, the U.S./allied air campaign that was designed to cripple Iraq's ability to wage war was estimated to have destroyed 40 percent of Iraq's tanks, 47 percent of its artillery, and tens of thousands of troops.[2] A decade later, during the war in Afghanistan, the Pentagon calculated that about 75 percent of the latest generation of U.S. bombs and missiles hit their targets, a significant increase over any previous war. In Afghanistan, precision-guided weapons—targeted by using satellites and lasers—

made up about 60 percent of the weapons, up from 8 percent during the Gulf War and 35 percent in the 1999 Kosovo war. The effectiveness of these weapons, said the Pentagon, was about 90 percent.[3] In Iraq in 2003, the highest tech weapons made up even a larger share of the U.S. arms inventory.

For the journalists covering war, this new precision changes the standards by which combat is judged. If weapons—primarily American weapons—can be targeted so accurately, then it should be possible to move back toward the kind of warfare in which combatants, not civilians, are the principal casualties.

This is the kind of issue that the news media should address. War coverage should be more than just an accounting exercise that adds up the numbers of people and objects destroyed. *Who* the victims are does matter, as does the strategy that determines who will be targeted.

—————

As the nature of conflict has changed, the ranks of those who wage war have expanded. Military historian John Keegan observed that "while war has become far too expensive, financially and emotionally, for rich states to wage with anything approaching the full potentiality, technological and human, their resources make available, it has also become, paradoxically, a cheap and deadly undertaking for poor states, for enemies of the state idea, and for factions in states falling apart. The rogue ruler, the terrorist and the fundamentalist movement, the ethnic or religious faction are all enemies as serious as any, in an age of junk weapons, as civilization has ever faced."[4]

Since fears of global war have receded, there is a tendency among policy makers, news organizations, and others who influence the public's perception of the world to view most modern warfare as a series of minor conflicts that have limited impact on anyone except those directly involved. These "small wars" attract little international attention because they occur in places that are outside the range of coverage by the most dominant—meaning Western—news media. Millions may die, as happened during the 1990s in Africa, without intruding into the global public's consciousness.

Ignoring these wars may be convenient, but it is irresponsible. Physical (and cultural) distance does not confer absolution on governments, the public, or the news media.

The War Boom

War continues to be a growth industry in much of the world. It is rooted in political and economic failure and in the absence of justice. Instigators' goals vary.

Mary Kaldor cites cases of war as being a form of political mobilization based on identity. "The military strategy," she says, "is population displacement and destabilization so as to get rid of those whose identity is different and to foment hatred and fear."[5] Rwanda shredding itself in 1994 stands out as an example of this kind of war.

In unhealthy states that are primarily collections of armed camps, war may be an appealing vocation. It is an outlet for anger and despair, and it is nourished by the world's most available commodity: weapons. Robert Kaplan reported that there may be 80 million firearms within Yemen—four for every Yemeni.[6] Little attention is paid to the devastating effect this embrace of war has on the future as well as the present. The economy, the culture, and virtually every other facet of life become hostages to conflict. When war arrives in a community, schools, medical clinics, and farms are abandoned and ignorance, illness, and hunger gain a lasting foothold.

For warriors with grand aspirations, the United States is an attractive target, symbolizing power, wealth, and arrogance. Beyond this general motivation, the reasoning of those who see America as an enemy often tends to be murky. Sometimes a specific rationale for an attack is cited, such as to punish the United States for its support of Israel. But generally it is America's economic and cultural heft that inspires animosity. The goals of attacks may be unclear. Strikes such as those on September 11, 2001, might spread fear and briefly satisfy a desire for vengeance, but they will not topple the American economy or government.

In the era of globalization, danger moves as freely as commodities, capital, and people, and that mobility increases any prospective target's vulnerability.[7] U.S. efforts to enhance homeland security show how difficult it is to secure a country, particularly if domestic freedom remains important. The question now for America's people and policy makers is how to balance that freedom with new realities of security. The question for the news media is how to devise coverage of all this in a way that incorporates long-term as well as immediate issues and helps the public understand the complexities of keeping the nation safe.

New Warriors and New Tactics

The relentless determination of today's most dangerous warriors is an important factor in understanding current incarnations of war. The Cold War was in many respects relatively passionless. There was intellectual fervor among ideologues on both sides and the two superpowers found each other politically despicable, but rarely to the point of frenzy. Among the pawns in the superpowers' game—nations such as Poland and the Baltic states—genuine anger smoldered, but only on rare occasions were people ready to shout, "Let us march!" (and mean it), particularly against a superpower.

Today, however, Holy War is resurgent. It brings with it a messianism overriding the rationality that survives as a trace element in most wars. Compared to Al-Qaeda, even groups such as the Irish Republican Army and the Palestine Liberation Organization seem conventional in their motives and persistence.[8] The privatization of terrorism, with terrorist organizations operating outside the structure of states, makes it difficult to rely on traditional measures such as deterrence and containment to counter them.[9]

The new terrorist organizations may have no definable homeland that can be designated as "enemy territory." Therefore the United States or other responding nations must be prepared to reach into host countries, sometimes with a volley of Cruise missiles and occasionally with a larger attack, such as the American expedition into Afghanistan in 2001. Because of territorial ambiguity, the insertion of U.S. special operations forces into even friendly states is likely to become more common.

Dealing with terrorists also requires that new attention be paid to legal and ethical issues such as assassination and torture. "Targeted killing" and use of all available means to extract information from prisoners may be unsavory, but an argument can be made that such tactics are legitimate forms of self-defense. As Middle East expert Steven Simon observed, when the enemy is intent on inflicting mass casualties, "unsavory may not be a sensible threshold."[10] Nevertheless, a strong case can be made that certain ethical boundaries must be respected and certain behavior must be not be endorsed. Whatever course is chosen, standards must be defined and limits set if a society is not to descend into moral anarchy.

Policies concerning such matters reflect the changing concept of war. This merits attention from the news media because the perceived legitimacy of the actors in conflicts depends partly on media depictions. News coverage can create heroes and villains, and can raise questions about behavior in direct and subtle ways. Although policy makers ultimately decide when and how to engage in conflict, journalists are also players in this process. Winning and keeping public support is crucial to those who govern, particularly when the rules of the game are changing. What seemed inappropriate yesterday may be necessary today, or after reconsideration it may be judged to still be unacceptable. If torture, assassination, and other previously shunned practices are now to be regarded as legitimate, the news media should stimulate thorough public debate about this. News organizations should provide print, broadcast, and on-line ven-

ues for exchanges between the public and policy makers about standards for behavior in postmodern conflict.

Society's principles are reflected and partly defined by the ways that war is
waged. New levels of violence might become necessary, but to simply stumble
into acceptance of them would be unconscionable.

Intervention and Mission

When armed intervention into another nation's affairs is being contemplated,
thoughtful, careful news coverage can help the public and policy makers alike
sort through options. The post–Cold War years have seen continuing debate
about intervention, and in this debate the tenets of international law and the
realities of a world of small wars do not always mesh. On some occasions when
intervention has been undertaken, it has been challenged by domestic constituencies and other nations. In other instances, the decision to *not* intervene
has been challenged.

There has been little consistency in standards or action. In 2003, the
United States cited human rights abuses in Iraq as one justification for war, although in 1988 America remained silent while Saddam Hussein used chemical
weapons against civilians. This kind of belated policy change infuriated some
human rights groups that denounced the revised American (and British) position as mere opportunism. The pragmatic view, however, as stated by international relations scholar Michael Ignatieff, is that "the fact that states are both
late and hypocritical in their adoption of human rights does not deprive them
of the right to use force to defend them."[11]

The 1999 conflict in Kosovo was another instance when the United States
and NATO did act, primarily through air attacks on Yugoslavia. Among the critics of this intervention was American lawyer Walter Rockler, who had been a
prosecutor at the Nuremberg war crimes trials. Noting the absence of a United
Nations mandate to act, Rockler asked, "Who gave us the authority to run the
world?" He added that "the attack on Yugoslavia constitutes the most brazen international aggression since the Nazis attacked Poland to prevent 'Polish atrocities' against Germans. The United States has discarded pretensions to
international legality and decency, and embarked on a course of raw imperialism run amok."[12]

Another viewpoint was offered by another American lawyer, Michael Glennon, who saw the Kosovo intervention as a precedent to which the international
legal and political systems would have to adjust. He wrote, "States will continue
to intervene, as NATO did in Kosovo, not where law tells them they may, but

where wisdom tells them they should, where power tells them they can, and—perhaps—where justice, as they see it, tells them that they must."[13]

Eloquent arguments can be mounted on both sides of the debate about intervention. The news media can help the public by fully covering not just the interventions themselves but also the longer-term questions behind the issues of the moment, such as what standards govern decisions about intervening and what precedents exist for current options. Many news organizations do not do this well; they tend to present news without context. Arriving late, journalists cover the aftermath of the explosion, not the causes leading to it and not in the timely way that might alert the world in time to snuff out the fuse. Flawed coverage has ripple effects; superficial reporting leads to superficial judgments about policy.

How individual nations redefine their role in helping to maintain international order is linked to even more complicated issues concerning the responsibilities and powers of organizations such as the United Nations and NATO. Again, the news media have an important job in defining issues, primarily to help the public understand the newest array of priorities and alliances.

Recent changes in NATO are examples of important measures that tend to be treated too casually by the news media. As the organization has grown, it has reconsidered its mission, looking beyond Europe. On September 12, 2001, NATO invoked Article 5 of the North Atlantic Treaty, which calls on all members to respond to an attack on another member as if they themselves had been attacked. This part of the treaty was written to provide the grounds for an American response to a Soviet attack on Western Europe, but it was brought into play after the United States itself was the victim of an attack.[14]

By late 2002, NATO was planning to establish a rapid deployment force of more than 20,000 troops that could operate anywhere in the world and that would include niche contributions from even the organization's smaller and newer members, such as Romania lending its mountain fighters. NATO had acted outside its boundaries in the Kosovo conflict of 1999 and in Macedonia in 2001, and individual members had aided the United States in the fighting in Afghanistan. By mid-2003, NATO had assumed command of the multinational force in Afghanistan. If NATO continues to expand its role and to create a permanent force for operations outside Europe, the entire global security structure could change, making the United Nations—which some see as too diverse and too slow to respond to crises—less relevant, particularly in military matters.

If past patterns hold true, among the last to grasp the significance of NATO's evolution will be the news media. The post–Cold War years are over, replaced by the post-9/11 period, and as labels of eras change so does the way the world works. News organizations, like governments, must adjust.

Of course, America remains dominant and that is unlikely to change anytime soon. Michael Ignatieff asked, "What word but 'empire' describes the awesome thing that America is becoming?" He points out that the United States is "the only nation that polices the world through five global military commands; maintains more than a million men and women at arms on four continents; deploys carrier battle groups on watch in every ocean; guarantees the survival of countries from Israel to South Korea; drives the wheels of global trade and commerce; and fills the hearts and minds of an entire planet with its dreams and desires."[15]

An American empire has long existed, regardless of reluctance to use the word. "Imperialism" is something ascribed to others, while Americans stick to their aw-shucks disclaimers about imperial intentions. That disingenuousness, always transparent, is increasingly pointless, and a growing number of commentators are adopting a more realistic approach to appraising American intentions. Robert Kagan cited the need to consider "the new reality of American hegemony."[16] Kagan said that Europe, at least, may not be inclined to challenge this status quo, because after a century of American intervention and protection, Europe is enjoying what appears to be a lasting peace. Its easiest path may be to accept America as the super superpower and let it do what it will. Although many Europeans may view Americans as intemperate cowboys, Europe, according to Kagan, is more likely to become irrelevant to the United States than to exercise a constraining influence on American power.

There are other views of what lies ahead for an imperial America. Charles Kupchan rated Europe's potency higher than Kagan did. In his book *The End of the American Era,* Kupchan wrote that relations between America and Europe might suffer not from criticism rooted in an inferiority complex, but because the new, peaceful, prosperous Europe will no longer feel dependent on the United States. He said, "Precisely because Europe is in the process of building a new political community that encompasses a region, not just a state, the EU may well find it convenient, if not necessary, to propagate a new and ambitious brand of pan-European nationalism."[17] Part of that might be seen in an EU military force that would be similar to NATO in some ways, but with one major difference—the absence of the United States, which despite its protests about an EU force undermining NATO, might be content to have even more leeway to act unilaterally.

Any estrangement or new rivalry between Europe and the United States that alters longstanding economic and military cooperation will represent a shift that may seem unnatural to many. With new concentrations of multilateral military power, the world will look different; whether better or worse may be hard to determine for a while. Amidst all the ponderous pronouncements at the beginning of this new century, little was said about how the flow of history can change. Predictability is elusive. The comfortable old slipper might be tossed out, replaced by a stiff new boot.

The politics surrounding the Iraq war illustrated this. Michael Ignatieff wrote: "As the Iraq debate at the United Nations showed so starkly, the international consensus that once provided America with coalitions of the willing when it used force has disappeared. There is no Soviet ogre to scare doubters into line. European allies are now serious economic rivals, and they are happy to conceal their absolute military dependence with obstreperously independent foreign policies. Throughout the third world, states fear Islamic political opposition even more than American disapproval and are disposed to appease their Islamic constituencies with anti-American poses whenever they can get away with it. . . . The days when the United States intervenes as the servant of the international community may be well and truly over. When it intervenes in the future, it will very likely go it alone and will do so essentially for itself."[18]

The news media should anticipate this kind of change and recognize the significance of debate about the responsibilities of power. Granted, faulty news coverage is not the only reason that most Americans have only the vaguest sense of the role of NATO or the UN and have no more than a tourist's understanding of any other country. The public has its own responsibility to maintain at least a baseline level of knowledge, and the nation's educational system has a similar duty to help build the intellectual foundation of global citizenship. Nevertheless, the news media contribute to collective ignorance through inadequate coverage. The American public's limited knowledge of European affairs is a good example of this problem because, from many Americans' standpoint, Europe is the least "foreign" part of the rest of the world. To contemplate the public's even greater lack of knowledge about Asia, Africa, Latin America, and elsewhere is to look into a bottomless chasm.

The Bush Doctrine

One of the key elements of changed American policy in the post-9/11 era is the broadening of military options and the increased willingness to employ substantial military force in pursuit of policy goals. During the Clinton administration, the grandest concept of conflict seemed to be air war, such as was employed in 1999 against Serbia.

That war raised its own strategic and moral questions for policy makers, the public, and the news media, given the virtual impunity with which the U.S./NATO forces could pummel their enemy:

- Does the ability to inflict but not incur casualties make war more palatable generally and more appealing as a policy option?
- Should the American news media simply report the glories of sanitized war, using dazzling graphics to depict the newest military hardware and its accomplishments and stressing the small number of U.S. casualties, or do journalists have a responsibility to foster public debate about all topics related to war?

The safety of American pilots is only part of the story. What happens on the ground—to enemy forces and to noncombatants—should also be reported thoroughly. Too often, however, chauvinism infects coverage and the perspective of "our side" is given disproportionate emphasis in reports presented to the public. Granted, the logistics of getting access to the other side may limit the breadth of stories, but news organizations should recognize the need to expand the scope of their reporting as far as possible.

After the 2001 attacks on the United States, George W. Bush's administration moved onto a level different from that of the Kosovo conflict. More comprehensive war was embraced as a policy tool. Part of the Bush approach was anticipatory self-defense, which he tersely summarized in a speech at the U.S. Military Academy in June 2002: "If we wait for threats to materialize, we will have waited too long."[19]

The Bush doctrine was more fully defined in the National Security Strategy presented by the White House in September 2002. The document cited the dangers "at the crossroads of radicalism and technology" and declared that "as a matter of common sense and self-defense, America will act against such emerging threats before they are fully formed." The new assertiveness that the Bush administration embraced was presented with straightforward justification: "History will judge harshly those who saw this coming danger but failed to act. In the new world we have entered, the only path to peace and security is the path of action."[20]

This Bush doctrine supersedes the Cold War strategy of containment, which blocked adversaries' expansionism and ultimately let authoritarian regimes, such as that of the Soviet Union, change (or rot) from within.[21] The new guiding principle was defined by Richard Haas, director of policy planning at the State Department, as "the limits of sovereignty." Sovereignty, he said, entails obligations: "One is not to massacre your own people. Another is not to support terrorism in any way. If a government fails to meet these obligations, then it forfeits some of the normal advantages of sovereignty, including the right to be left alone inside your own territory. Other governments, including the United States, gain the right to intervene." That broad premise becomes more precise concerning terrorism, which, said Haas, can "lead to a right of preventive, or peremptory, self-defense. You essentially can act in anticipation if you have grounds to think it's a question of when, and not if, you're going to be attacked."[22]

At the heart of this change is not a new bellicosity, but rather a recognition that the technology of war has once again changed the politics of war. Terrorism is no longer a matter of planting a packet of explosives along a city street or in an airplane, horrific as the results of those acts might be. Terrorism now might involve weapons of mass destruction. What was once the scenario of Tom Clancy novels has become a real threat. Terrorism's menace also puts nations such as Afghanistan on the list of states that the United States views as threatening to American security. Remoteness does not necessarily remove the threat. Michael Ignatieff noted that "terror has collapsed distance, and with this collapse has come a sharpened American focus on the necessity of bringing order to the frontier zones."[23]

The changes that result from this can be difficult to grasp. In the debate during early 2003 about dealing with Iraq, some of the argument against prompt, forceful action was grounded in outdated assumptions about Saddam Hussein's capabilities. If he—or others like him—were posing a threat in conventional terms, such as by massing infantry and tanks that he might send into a neighboring country, then he could be dealt with in conventional ways. Deterrence presumably would suffice; otherwise, punishment such as that inflicted on Iraq in 1991 could discourage further adventurism.

That approach is outdated. If a state or a non-state terrorist organization possesses biological, chemical, or nuclear weapons, the stakes become very different. Traditional measures such as deterrence or containment may be too risky because of the relative ease and speed with which those weapons can be used against a puny neighbor or a superpower. This kind of thinking was at the heart of the most forceful arguments in support of invading Iraq.[24]

International affairs scholar Joseph Nye said that the Bush national security strategy "makes a plausible general argument for preventive war," given that terrorists being armed with weapons of mass destruction would be "not only a major change in world politics; its potential impact on our cities could drastically alter our civilization." That is a frightening concept that news coverage has helped make comprehensible through reports not only about 9/11, but also about terrorist attacks elsewhere, although none of these have been of the magnitude that terrorists might eventually launch.

Much more challenging is addressing this phenomenon within the context of a world order that has come to rely on measured, multilateral responses to threats to peace. Nye pointed out that systemic responses to such threats could be designed within the scope of the United Nations Charter.[25] And, of course, however the response is designed, evidence of the existence of the threat must be solid enough to justify extraordinary action such as an invasion.

The debate about the wisdom of acting against Iraq will continue, partly because of the prospect of that war serving as precedent for further unilateral intervention. International law scholar Richard Falk was among the critics of the Bush approach. "It is a doctrine without limits," he wrote, "without accountability to the U.N. or international law, without any dependence on a collective judgment of responsible governments and, what is worse, without any convincing demonstration of practical necessity."

Falk agreed that "the reality of the mega-terrorist challenge requires some rethinking of the relevance of rules and restraints based on conflict in a world of territorial states. The most radical aspects of the Al-Qaeda challenge are a result of its nonterritorial, concealed organizational reality as a multistate network. Modern geopolitics was framed to cope with conflict and relations among sovereign states; the capacity of a network with modest resources to attack and wage a devastating type of war against the most powerful state does require acknowledgment that postmodern geopolitics needs a different structure of security."

Nevertheless, wrote Falk, "what is needed is new thinking that sees the United States as part of a global community that is seeking appropriate ways to restore security and confidence, but builds on existing frameworks of legal restraints and works toward a more robust U.N. while not claiming for itself an imperial role to make up the rules of world politics as it goes along."[26]

In the coming years, policy makers will consider issues related to preemptive war and the unilateralism versus multilateralism debate. These are among the issues that will shape global politics and influence decisions about war and peace. The news media could do more to help the public understand these matters. Even following the American occupation of Iraq, such topics were not adequately debated, and within the United States unilateral American intervention seemed to be tacitly accepted as part of the new order of things.

Discussions about intervention keep coming back to the issue of the presence—proven or suspected—of weapons of mass destruction, which can be the great equalizer. As the Bush national security strategy says, it would be folly to allow continued existence of this kind of power when coupled with credible threat. But given the proliferation of advanced weapons technology, this principle could become the springboard for any number of wars in the near future. Policy makers face the task of dealing with different kinds of threats and explaining the distinctions among the threats to the public. Why attack Iraq but allow provocative behavior by North Korea and Iran? At what point should India and Pakistan be considered to be threats to more than each other? The list of questions will grow and the answers will do much to define the state of the world in the coming years. Journalists and the public should press policy makers to at least begin developing those answers.

Covering New War

The U.S. audience for war coverage is not sure what it wants from news organizations. Americans say that a free press is important and that the news media should question the government about its policies. But those are peacetime opinions, and attitudes shift considerably during wartime.

Survey research conducted in January 2003 for ABC News asked what was more important during wartime, the right to a free press or the government's ability to keep secrets. By 60 to 34 percent, respondents favored the government's keeping secrets. When asked whether the news media should support or question the government's war effort, 56 percent said support, 36 percent said question.[27]

Such popular sentiment makes journalists' work more difficult. It is hard to serve the public if the public doesn't like the way it is being served. So during wartime the news media can expect plenty of criticism if they are perceived as being less than supportive of the war effort. "Less than supportive" may mean reporting bad news from the battlefield or pointing out apparent flaws in war policy. The coverage might be totally accurate, but that won't matter to those whose longstanding anti-media feelings are boosted by a dose of patriotic fervor.

In such an environment, journalists must be especially careful not to be intimidated by government officials or public opinion. News executives, concerned about keeping their audience, read the polls just like politicians do and sometimes forsake aggressive news gathering while trying to conform to the public mood of the moment.

War coverage has always involved considerable tension between the press and the military. The erroneous but still widely held notion that news coverage was a key factor in the loss of the Vietnam War led to increasingly restrictive coverage rules imposed by the Pentagon in conflicts such as the Persian Gulf War of 1991. Over the longer term, concern bordering on psychosis has influenced the military's policies related to news coverage.

In addition to political concerns, worries about operational security (some reasonable, some not) and logistics have led military planners to try to limit news gathering and reporting. Journalists, meanwhile, have acquired tools such as satellite videophones that facilitate real-time coverage from even the most remote battlefield. To avoid a head-on collision with the news media, the Pentagon devised the embedded journalist program for use in Iraq, which is discussed in chapter three. Department of Defense officials also employed more creative ways to sidestep traditional news coverage. They agreed to give movie producer Jerry Bruckheimer special access to American troops in Afghanistan in 2002 to create a "reality" television series, *Profiles from the Front Line*, that would be aired during prime time by ABC's entertainment division. When ABC News executives complained to officials at the parent company, Walt Disney, they were rebuffed. As part of the deal, the Pentagon would review Bruckheimer's footage and ABC would make any changes that were requested "for the safety of the troops." ABC's news division would not have early access to the film, regardless of its news value.[28]

The news media's integrity is more than just an internal matter for the journalism profession to address. News coverage does much to establish the political framework within which war is conducted. Writing about the coverage of the 1991 Gulf war, *New York Times* correspondent Chris Hedges said: "It gave us media-manufactured heroes and a heady pride in our military superiority and technology. It made war fun. . . . Pool reporters, those guided around in groups by the military, wrote about 'our boys' eating packaged army food, practicing for chemical weapons attacks, and bathing out of buckets in the desert. It was war as spectacle, war as entertainment. The images and stories were designed to make us feel good about our nation, about ourselves. The Iraqi families and soldiers being blown to bits by huge iron fragmentation bombs just over the border in Iraq were faceless and nameless phantoms."

Hedges continued, "The notion that the press was used in the war is incorrect. The press wanted to be used. It saw itself as part of the war effort. . . . For we not only believe the myth of war and feed recklessly off the drug but also embrace the cause. We may do it with more skepticism. We certainly expose more lies and misconceptions. But we believe. We all believe."[29]

If Hedges is correct—and he has covered many wars—objectivity in war reporting is skewed from the start, distorted by boosterism that is dressed up as "patriotism." Despite pretensions of detachment, journalists are actually part of the machinery of war. It is easy and very appealing to "get on the team" even while professing judicious independence, but the news product will reflect this. Sometimes even those in the news business are slow to recognize that patriotism does not require backing down from truth.

The Content of War Coverage

News coverage is more than the presentation of facts. It includes a substantial dose of opinion, and this is not limited to the clearly identified homes for opinion such as the editorial page.

A study of American television news in the months immediately following the 2001 attacks on the United States found high levels of opinion ("punditry") coupled with uncritical reporting of U.S. government statements. The study, conducted by the Project for Excellence in Journalism, found that newspaper coverage was more fact-based than was television news, and that television news was measurably less likely to include criticism of the government than were the print media.[30] By nearly an 8:1 ratio, the stories examined in the study reflected support for U.S. actions rather than dissent. When coverage is so lopsided, the public's perception of events is likely to be narrow. Intellectual diversity may be squeezed out.

Parochialism, however, has its rewards. The study found that during those days shortly after 9/11 69 percent of Americans believed that journalists "stand up for America," compared to 43 percent a few months earlier, and 60 percent believed that the press "protects democracy," up from 46 percent.[31]

News executives might bask in this rare approval, but while everyone is feeling patriotically warm and fuzzy the journalistic product suffers. Certainly, the people who run news organizations are concerned about circulation and ratings, but over the long run they might find that courting popularity is more costly than resisting the emotions of the moment. Eventually, the public might realize the value of objective presentation of the news and feel deceived if they have been getting something else.

Even absent a major crisis, news coverage inevitably includes bias of one kind or another. This is not necessarily overt ideological favoritism. More common are subtle prejudices, born of life experience, that slip into news stories and decisions about what to cover. But even if no political agenda is involved, such tilting may have effect. Sometimes, slanting of the news may be rooted in semantics, as can be seen in coverage of terrorism.

The word "terrorism" signifies evil. But words matter and so does the way in which the news media describe acts of terrorism and those who commit them. Terrorism needs explaining; it is more than mindless fury that has no comprehensible rationale behind it. For some, it is a legitimate tool of war, just as a battalion of tanks is, to be used when it appears to be the best weapon available. The troops of terrorism, suicidal or not, may be deployed as part of a larger strategy with defined goals. To dismiss all acts and perpetrators of terrorism as irrational is to underestimate them, and when news coverage relies on this characterization, it may mislead the public.

In their coverage of terrorism (as with other topics) journalists should do more than conform to the dominant political position of the moment and accommodate current public sentiment. The public deserves to be able to examine issues from varied perspectives. In the mainstream media, rare are voices such as that of writer Susan Sontag, who argued in *The New Yorker* immediately after the September 11 attacks that those "licensed to follow the event seem to have joined together in a campaign to infantilize the public. Where is the acknowledgement that this was not a 'cowardly' attack on 'civilization' or 'liberty' or 'humanity' or 'the free world,' but an attack on the world's self-proclaimed superpower, undertaken as a consequence of specific American alliances and actions?"[32]

Sontag did not adopt the politically convenient formula that goes no farther than defining perpetrators of terrorism as bad and their victims as good. Her approach may anger many, but it may also cause some readers to think, not merely react. Considering why an act of terror happened is important, and should not be ignored by news organizations intent on reducing the story to a showcase for knee-jerk patriotism.

Communications scholar Bethami Dobkin noted that television news, in particular, "complements a political process that is reliant on public images for legitimacy and guidance." Dobkin also said that television's heavy emphasis on images, as opposed to the political ideas behind the images, makes terrorism more incomprehensible.[33] The images of the World Trade Center as it was hit by the airplanes and as it crumbled shortly thereafter are so vivid and disconcerting that they dominate public perception of the event. It is the responsibility of the news media to expand the public's focus and tell the larger story of why the act happened. Memory will retain images of the attack, but they should be supplemented by reporting that establishes context.

Labeling the action and the perpetrators is another issue that journalists must address. Media and terrorism scholar Brigitte Nacos wrote that the mass media, recognizing the various views and negative connotations of the terms "terrorism" and "terrorist," "seem uncertain and confused as to when to describe political violence as terrorism and when to choose other labels. Often, reporters and editors tend to take their cue from government officials in this respect. The result is an inconsistent use of several terms describing the perpetrators of terror . . . and their deeds. "[34]

Shortly after the September 11 attacks on the United States, Reuters News Service announced it would use the word "hijackers" rather than "terrorists" to identify those who seized the airplanes and used them as weapons. Reuters's explanation was that one person's "terrorist" is another person's "freedom fighter." That decision was strongly criticized by those who saw it as pandering to anti-American sentiment.

The labeling issue rose again when Chechens seized a Moscow theater and more than 700 hostages in October 2002. News organizations used various labels to describe the perpetrators. American news reports referred to "gunmen," "hostage-takers," "guerrillas," "rebels," and other such terms, but only occasionally "terrorists."[35] This drew criticism from Vasily Bubnov in *Pravda,* who asked, "Why does the Western press refer to Chechen terrorists as rebels?" and said: "It is obvious that they are not freedom fighters. They are simply bandits."[36]

According to Bethami Dobkin, in television news, with its tendency to use visual and spoken shorthand, "description often substitutes for explanation. What we often overlook is that description implies an attitude, a point of view. Television relies on stereotypes for the sake of simplicity or ease of depiction, providing us with, for instance, portrayals of Arabs as crazy, suicidal, religious fanatics. These stereotypes spill over from the terrorists, informing our judgments about all Arabs. . . ."[37] CBS anchorman Dan Rather, while appearing on *The Late Show with David Letterman* shortly after the September 11 attacks, used that kind of shorthand: "They hate America. . . . They want to kill us and destroy us. Who can explain madmen and who can explain evil? They see themselves as the world's losers. . . ."[38]

Granted, Rather did not say this on his CBS newscast, but if allowed to affect news coverage, the attitudes behind sweeping statements such as Rather's will pull journalism farther away from objectivity. In most cases, it *is* possible to "explain madmen" and "explain evil." That is journalists' job. Terrorism requires sophisticated responses from the news media as well as governments and the public, and not simplistic, formulaic approaches.

Some news organizations have addressed this issue by thinking about the meaning and impact of words and trying to maintain consistency in the use of those words. The style guide of the British newspaper *The Independent* takes this approach: "Terrorism is a violent action intended to create terror among a civilian population so as to destabilize a government. Thus, an IRA man who plants a bomb in a public house is acting as a terrorist; one who shoots a British soldier is not. Useful words for civilians who take up arms against a government are revolutionary, militiaman/woman, rebel, and paramilitary; their actions may make them bombers, gunmen, hijackers, or killers. Guerrilla and insurgent are useful because they carry no charge of condemnation or approval. . . . Avoid loaded phrases such as resistance fighter and freedom fighter unless in direct speech."

Writing about this semantics issue, Guy Keleny of *The Independent* said that it "sounds fine in theory, but I am not wholly convinced. It seems to require an insight into the motivations of the 'terrorists' that will rarely in practice be tolerable." He noted that the paper's Middle East staff had reported that "Palestinian militants" had sent another suicide bomber to kill Israelis. He said that the reporters "do not use 'terrorist' outside quotation marks, because it is the favored language of Israeli propaganda. The same applies to the favored language of Palestinian propaganda, and 'martyr' is not used naked either."[39]

Tim McNulty of the *Chicago Tribune* said that at that newspaper, "terrorism" is linked to a deed, not an organization. "In the '80s and earlier and into the '90s," said McNulty, "the Israelis called the PLO a terrorist organization. Then they became their negotiating partner, and the term 'terrorist organization' dropped out of the lexicon. In Beirut, there were social workers and teachers who were members of the PLO, and to label everyone a terrorist seemed absurd. In our paper we try to keep that term associated with an act rather than with a group of people or supporters of that group of people."[40]

McNulty's point underscores the dangers of casual labeling. As he notes, some people associated with the PLO may be terrorists, but that does not mean that all PLO members are terrorists. Care is required even when using that guideline. International affairs scholar Walter Laqueur said that "to call a terrorist an 'activist' or a 'militant' is to blot out the dividing line between a suicide bomber and the active member of a trade union or a political party or club. It is bound to lead to constant misunderstanding."[41]

Washington Post guidelines reflect similar concerns, saying, in part: "The language we use should be chosen for its ability to inform readers. Terrorism and terrorist can be useful words, but they are labels. Like all labels, they do not convey much hard information. . . . When we use these labels, we should do so in ways that are not tendentious. For example, we should not resolve the argument over whether Hamas is a terrorist organization, or a political organization that condones violence, or something else, by slapping a label on Hamas. Instead we should give readers facts and perhaps quotes from disputing parties about how best to characterize the organization. . . . We should always strive to satisfy our own standards and not let others set standards for us."[42]

Imprecise labels take on lives of their own, infecting public discourse. In the minds of many Americans, the logical word following the modifier "Islamic" is "terrorist." That misconception is not entirely the fault of the news media, but news coverage may be a contributing factor if it fails to avoid sloppy description.

The Canadian Broadcasting Company also proceeds carefully, describing Palestinian perpetrators of attacks on Israelis as "militants," "gunmen," and "assailants," but not "terrorists." CBC ombudsman David Bazay said in 2002 that speaking of "Palestinian terrorism" would be asking the public broadcaster "to take sides and to embrace the Israeli government's position and its definition of terrorism, which denies the legitimacy of Palestinian resistance."

But that does not mean that an act of terror should receive gentle semantic treatment. In a memo for the CBC staff, Bazay said that "militant" is a "generic expression referring to all the Palestinians who are actively resisting Israeli occupation," while "terrorist" should be used only after the fact of an attack. In other words, said Bazay, nothing should keep CBC journalists from calling "a terror attack a terror attack."[43]

Despite efforts to impose clarity, the news media, like politicians, sometimes further simplify issues by using one person to symbolize a complex topic. Demonizing a "master terrorist," captures the public's interest and is easily understood. After the 2001 attacks on the United States, the "war on terrorism" was portrayed as primarily the hunt for Osama bin Laden. According to Brigitte Nacos, "the news media's tireless focus on 'the world's most wanted terrorist' seemed to attribute the terror problem to one star terrorist and his group, fostering a false perception as to the scope of the threat and the prospect for its removal."[44]

This approach by the news media is deeply flawed. Placing so much emphasis on an individual oversimplifies and distorts the dynamics of conflict. Attributing the attack on the United States to a single criminal mastermind might work well in a Hollywood screenplay, but it is flawed journalism. Osama bin Laden and other individuals represent complex forces that transcend personalities. Those forces, not just their "star players," should be covered thoroughly. Even Al-Qaeda has been simplistically portrayed in many news reports as a cohesive organization that is engineering a worldwide terror campaign, while in truth Al-Qaeda might be a much more scattered and loosely constructed entity (which could make it even more frightening). Sometimes governments cite doubtful ties to Al-Qaeda to demonize politically undesirable individuals and groups.[45] They may also be hoping to tap into American anti-terrorism funding.

A related task in the coverage of terrorism is discriminating between news and speculation, and deciding when it is appropriate to publish the latter. Michael Getler, ombudsman of *The Washington Post,* wrote about a story that illustrated the problems accompanying such a decision. The story in question ran in *The Post* in December 2002 and was headlined "U.S. Suspects Al-Qaeda Got Nerve Agent From Iraqis." The lead was, "The Bush administration has received a credible report that Islamic extremists affiliated with Al-Qaeda took possession of a chemical weapon in Iraq last month or late in October, according to two officials with firsthand knowledge of the report and its source."

That sounds like an important story. But in the story were these passages: "If the report proves true"; "not backed by definitive evidence"; "the principal source on the transfer was uncorroborated"; "open to interpretation"; and others. As Getler noted, *The Post* did the right thing by letting readers know about all this uncertainty, but the question remained: Should the story have run, or should it have been held until better information was available?

Managing editor Steve Coll said the story deserved publication because the report had led to a "robust, if largely secret response within the national security establishment" and the public needed to know about that. Also, the story had been tracked down by *The Post*'s own reporter, not handed to the press by the administration as so much of this kind of information is.[46]

Trade-offs are unavoidable, and in this case *The Post* handled them properly. There was enough information about the government's response to be worth a story, and the paper made it clear that there were questions about the reported deal between Al-Qaeda and Iraq. In such cases, the public deserves to be told, but then the journalists need to keep working on the story and follow up with a more complete version when more information is assembled.

Different Perspectives

Every nation's journalism has its own character. Some countries' news media are aggressively independent, while others stay close to their government's official line. Some pride themselves on high standards of accuracy, while others are more free-wheeling.

Even when countries have much in common culturally, as is the case with the United States and the United Kingdom, journalistic approaches may reflect substantive differences. The fierce competition among British news organizations, particularly London's ten daily papers, fosters aggressive reporting with a political edge that is uncommon in much of mainstream American journalism. George Brock, managing editor of London's *Times,* cited the tradition of partisanship and polemical writing in Britain, and said that "intelligent readers understand that an artificial requirement for 'balance' hampers, rather than promotes, understanding."[47]

Differences between U.S. and U.K. media were apparent after the attacks on the United States in 2001. British news media reported the September 11 attacks extensively and sympathetically, but during the following months treated the war in Afghanistan differently. Stories about U.S. strikes that resulted in civilian casualties received more coverage from the British news organizations, which were more skeptical than their American counterparts about official statements that downplayed such incidents.

Beyond the example of British coverage, different judgments about newsworthiness were apparent in international reporting about treatment of Afghan prisoners by American forces, alleged U.S. spying at the United Nations, and possible forging of documents that were used as evidence of Iraq's misbehavior. These stories were all treated more casually by American media than by news organizations elsewhere. Most of the U.S. coverage seemed grounded in the belief that American policy was benign, if not always wise, while international coverage reflected greater doubt about U.S. actions.[48]

Nothing in this is particularly surprising, but it is important because the tone as well as substance of news coverage has ripple effects. Americans who wonder about the vilification of the United States several months later during the war against Iraq should recognize the continuing role the international media play in shaping, directly or indirectly, the intellectual environment in which anti-American opinion flourishes. (International reporting about the Iraq war is examined in chapter six.)

Aside from the effect of news coverage on the public's opinion about specific events, media representations contribute to general attitudes about war. Photographs are good examples of this, shaping perceptions of what war is like and evoking strong emotional responses to the horror, courage, and other elements of war.

These responses may vary greatly. At the time of the Spanish Civil War, Virginia Woolf wrote in *Three Guineas* that the reactions of men and women to images of war are distinct because men find "some glory, some necessity, some satisfaction in fighting," while women do not. But perhaps, she wrote, men and women might someday agree that "war is an abomination; a barbarity; war must be stopped."[49]

Susan Sontag cited Woolf in her own examination of the power of images. Sontag wrote that horrific photographs might "vivify the condemnation of war, and may bring home, for a spell, a portion of its reality to those who have no experience of war at all." But, she added, "someone who accepts that in the world as currently divided, war can become inevitable, and even just, might reply that the photographs supply no evidence, none at all, for renouncing war—except to those for whom the notions of valor and of sacrifice have been emptied of meaning and credibility. . . . Photographs of an atrocity may give rise to opposing responses. A call for peace. A cry for revenge. Or simply the bemused awareness, continually restocked by photographic information that terrible things happen."[50]

She wrote that "a good deal of stoicism is needed to get through the newspaper each morning," and she urged people to also remember the pictures, cruelties, and deaths that are *not* being shown.[51]

For journalists the issue remains, is the coverage—pictures and words— making war real to those who rely on the news media to tell them the truth about what is happening? Certainly, people will decide about what is "real" according to a variety of criteria, such as their own personal experience and knowledge. Gender may make a difference, as Woolf suggests, as might other factors.

When Woolf wrote about pictures from the Spanish Civil War, photographs such as those she cited were not common fare in mainstream publications. Today, still and moving images related to conflict flash incessantly across television and computer screens. This frequency may have a sensitizing or a desensitizing effect, depending on who is watching. The video of the airplanes crashing into the World Trade Center is at first horrifying. But after the images have been seen dozens or even hundreds of times, they may lose their edge, becoming "just television." Similarly, after seeing so many terrible pictures from wars around the globe, many of them presented as freestanding "news" without context, what is the public supposed to think about the state of the world? Do such images inspire people to respond or merely to despair, or even to dismiss the events being depicted as not relevant to their own lives?

The ability to send and receive such images continues to expand, as photography and communications technologies advance. In wartime, real-time coverage from the battlefield is becoming commonplace, and plenty of Web sites, as well as television, will bring us such coverage non-stop. Maybe that's good. Perhaps if there had been Webcams scattered throughout Rwanda in 1994 and if pictures of the genocide as it happened had been transmitted to the world, someone would have acted. Or maybe what was occurring would have been viewed as just one more "reality" program, watched with practiced neutrality for a short while until boredom set in and viewers switched to something else.

During the early days of the Iraq war, a number of Middle Eastern news organizations showed graphic photographs of what were said to be soldiers and civilians killed in the fighting and captured American soldiers being interrogated. American news organizations showed only brief glimpses of those images. *The Washington Post*'s Philip Kennicott asked: "Are images facts or illustrations? If a fact is ugly, should it be kept at a distance from readers and viewers? And what do news organizations do with the simple fact that there is both an eager appetite for, and a sincere disgust with, graphic images?" Kennicott said that to many American news organizations "the images were more a matter of illustration, something supplemental and discretionary that wasn't necessary to fully cover the capture of American soldiers. The *fact* was the capture and possible execution of some POWs; the images were a graphic addendum. And so most American news organizations chose to keep viewers at one descriptive remove: They would tell viewers what was on the tape, but not show it."[52]

This issue also arose on ABC, when *Good Morning America* host Charles Gibson said, "Anytime you show dead bodies, it is simply disrespectful, in my opinion." *Nightline*'s Ted Koppel (who was with U.S. forces in Iraq) disagreed, saying, "I feel we do have an obligation to remind people in the most graphic way that war is a dreadful thing."[53]

If the news media's job is to report war as it is, not sanitize it, then news organizations should deliver undiminished reality to their audiences. The public, and certainly the government, may not like it, but there's nothing wrong with that. If the news media present war as neat and tolerable, then the public is being deceived and attitudes about war will be distorted. Images in themselves can be facts, not mere supplements to words, and should be presented as such.

Chris Hedges has stressed the importance of using the words and pictures that best convey war's reality. "War itself," writes Hedges, "is venal, dirty, confusing and perhaps the most potent narcotic invented by humankind. Modern industrial warfare means that most of those who are killed never see their attackers. There is nothing glorious or gallant about it. If we saw what wounds did to bodies, how killing is far more like butchering an animal than the clean and neat Hollywood deaths on the screen, it would turn our stomachs. If we saw how war turns young people into intoxicated killers, how it gives soldiers a license to destroy not only things but other human beings, and if we saw the perverse thrill such destruction brings, we would be horrified and frightened."[54]

The public must decide for itself how to deal with reality. Writing about Germans' contemplation of their violent past, German writer W. G. Sebald said, "We are always looking and looking away at the same time."[55]

Such ambivalence is also part of the news media's approach to war. Even when journalists look at the results of war—the dead child, the burned-out village, the huddled refugees—when it comes time to confront the reasons behind the agony, reasons that require careful and perhaps painful explanation, some news organizations turn away.

———

Despite the long history of war reporting, questions continue to be raised about graphic reality and softened depictions, objectivity and patriotism, and many more issues that affect the quality of coverage. During the Iraq war the news media would find their work shaped by new technology, new procedures, and, for many, a new relationship with the military. But they would still keep looking for answers to basic questions about what war journalism should be.

CHAPTER THREE

Technology and Empathy:

The New War Journalism

When the United States goes to war, the news media's job is to provide extensive, independent coverage, whether the government likes it or not. In a democracy that's the way it's supposed to be. But the niceties of democratic theory don't always bother the many government officials who try to limit and even dictate the coverage of conflict.

They may cite "national security" as the reason for such constraints, but they are often more concerned about political security. The public can be slow to distinguish between the two and tends to side with government flag-wavers, at least for a time. Reining in the news media becomes the patriotic thing to do.

Even when journalists try to evade government obstruction, the logistics of war coverage can work against them. Television reporting in particular long required a cumbersome electronic support system, and print reporters also need ways to transmit their stories. Because the military was able to control much of the access to and communication from the battlefield, coverage could be choked off.

Within the past few years, that has changed dramatically. Increasingly mobile and decreasingly costly communications tools give journalists unprecedented range and independence. Carrying their own gear with them, they can bounce words and pictures off satellites and provide real-time coverage from almost anywhere. Also, the international ranks of the news media have expanded, meaning a larger and more diverse corps of journalists is reporting from war zones.

This proliferation of journalists and high-tech hardware has forced governments to rethink how they might influence coverage. In the Iraq war, these changes altered the press–military relationship and gave rise to a new kind of war journalism. But first, some not-too-ancient history.

Who Lost Vietnam?

It is an undying myth: News coverage was a major reason that the United States lost the Vietnam war. This notion is grounded in invention, not fact. It is a flimsy excuse for a failure rooted in flawed policy, but it remains potent, especially in military and political circles. In essence, the argument is as follows: American news organizations provided inaccurate and biased coverage of events in Vietnam, overstating the accomplishments of communist forces and putting negative spin on American policy makers' reasons for fighting. As a result, political support for the war disintegrated, making the military effort unsustainable.

Even myths may have a sliver of truth in them. As with any complex story, the reporting was far from perfect. Mistakes were made, such as in the coverage of the 1968 Tet offensive, which initially portrayed what was a significant military setback for the communist forces as a success. In his detailed study of the Tet coverage, journalist Peter Braestrup said the reporting was "a distortion of reality—through sins of omission and commission—on a scale that helped shape Tet's repercussions in political Washington and the Administration's response."[1] That became accepted wisdom among those who fought in Vietnam and became commanders in later wars. Colin Powell is one of these, and he wrote in his memoirs that "the images beamed into American living rooms of a once faceless enemy suddenly popping up in the middle of South Vietnam's capital had a profound effect on public opinion. Tet marked a turning point, raising doubts in the minds of moderate Americans, not just hippies and campus radicals, about the worth of this conflict, and the antiwar movement intensified."[2]

Lyndon Johnson learned this painful lesson. After leaving the presidency, he said about the Tet coverage, "The media seemed to be in competition as to who could provide the most lurid and depressing accounts." He added that the "daily barrage of bleakness and near panic" in the news reports convinced the public that Tet had been a defeat for the United States and its South Vietnamese allies.[3]

But a major reason for the negative tone of the Tet reporting and other coverage of Vietnam was not that journalists were determined to undermine the war effort, but rather that they—and the public—had begun to realize that they had been too accepting of the U.S. government's inflated appraisals of progress. Clark Clifford, secretary of defense at the end of Johnson's presidency, wrote in his memoirs that "reporters and the antiwar movement did not defeat America in Vietnam. Our policy failed because it was based on false premises and false promises. Had the results in Vietnam approached, even remotely, what Washington and Saigon had publicly predicted for many years, the American people would have continued to support their government."[4]

Some of the flaws seen in coverage of Vietnam persist today. Kevin Buckley, who was a reporter in *Newsweek*'s Saigon bureau during the war, said that "there were two big questions: How are we doing and what are we doing?" Focusing on "how are we doing" may leave unanswered the more important questions about what is being done. Buckley said that in Vietnam, "a government official would say, 'Ninety-two percent of that province is pacified.' A reporter might take the official to task about the 92 percent without asking the real question, 'What is pacification?'"[5]

The domestic repercussions of the Vietnam war still shadow American political life, even if moral issues about that war are less well remembered. Johnson's successors in the White House have subscribed to his view of media influence. They have taken note that this president who had won more than 60 percent of the popular vote in 1964 felt that he could not run again four years later largely because of press-fueled public sentiment about the war. With that in mind, when recent presidents have sent U.S. forces into combat, they have taken steps to prevent the news media from causing them the same kind of political harm.

From Grenada to the Persian Gulf

When it seized Grenada in 1983, the Reagan administration didn't bother with subtlety in its policy toward news coverage. Secretary of State George Shultz stated the administration's view: that journalists are "always seeking to report something that's going to screw things up."[6] No reporters were allowed to accompany American troops and those who tried to get to the scene on their own by boat were intercepted by the U.S. Navy.

The previous year, British officials had set an example for their American cousins by tightly limiting news coverage of the Falklands war. British journalist Robert Harris wrote that "the Vietnam analogy was a spectre constantly stalking the Falklands decision makers and was invoked privately by the military as an object lesson in how not to deal with the media."[7]

A principal government tactic was delay. Some television reports took twenty-three days to get from the Falklands to British newscasts. By contrast, during the Crimean war in 1854, the charge of the Light Brigade was described graphically in London newspapers twenty days after it happened. Falkland news reports had to be "cleared"—meaning censored—by British authorities, and when angry journalists tried to mention this in their stories they found that the word "censored" had itself been censored.[8] Overall, wrote British journalist Ian Trethowan, political leaders' eagerness to restrict coverage "showed an alarming lack of confidence in the emotional sturdiness of the public on whose behalf they claim to govern."[9]

American news organizations' unhappiness about the barriers to coverage of the Grenada incursion led to the creation of a panel chaired by General Winant Sidle that recommended using a media pool—a rotating contingent of journalists that would accompany U.S. forces during the earliest hours of military action before operational security allowed open coverage to begin. During the 1989 invasion of Panama, the pool ended up being sequestered at a military base and was unable to begin its coverage until the second day of fighting. (The pool members watched the invasion on CNN, which had a Panama City bureau.) When large numbers of journalists arrived, they were kept at Howard Air Force Base, where 850 members of the press corps had only two phones (neither of which worked well) available for filing stories.[10]

Joint chiefs chairman Colin Powell acknowledged problems in the media pool system and in a May 1990 directive reminded commanders that "otherwise successful operations are not total successes unless the media aspects are properly handled. . . . The media aspects of military operations are important, will get national and international attention, and warrant your personal attention."[11]

A postwar study of the 1991 Gulf war found that the news media's coverage had evolved "without much planning or reflection and certainly with little historical perspective on the role of the press in wartime. At the same time, the military had made studious preparations for dealing with the press in this war, far beyond what it had ever done before."[12] Recognizing that an attempt to censor reporting would stir up more controversy than it was worth, the Pentagon instead took advantage of the commitment by CNN and others to provide massive amounts of coverage. Defense Department planners knew that networks providing such coverage would be desperate for enough material to fill air time and would indiscriminately accept most of whatever was offered to them. Every day, U.S. military officials held lengthy briefings, many of which were carried live, and provided an ample supply of video, such as from bomb-mounted cameras. For the Defense Department, this was the best of all worlds: live briefing coverage brought the Pentagon's version of events directly to the public without much journalistic editing or commentary, and the bomb videos were impressive and sanitized depictions of American might.

The New York Times reported that the goal of senior administration officials, including the president, was "to manage the information flow in a way that supported the operation's political goals and avoided the perceived mistakes of Vietnam."[13] Late in the war, President Bush himself remarked that defeatism was gone—"By God, we've kicked the Vietnam syndrome once and for all."

Administration efforts to control coverage had several things going for them. The realities of live television reporting from the war zone raised issues of operational security, since members of the Iraqi high command were presumably among the audience for CNN and other networks. Also, American public opinion was solidly behind restrictions on the press. A Times Mirror poll conducted during the Gulf war's first month found that 79 percent of respondents thought military censorship was a good thing, and 57 percent favored even tighter military control of the news media.[14]

Among the journalists who objected to government constraints was Walter Cronkite, who said: "An American citizen is entitled to ask, 'What are they trying to hide?' The answer might be casualties from shelling, collapsing morale, disaffection, insurrection, incompetent officers, poorly trained troops, malfunctioning equipment, widespread illness—who knows? But the fact that we don't know, the fact that the military feels there is *something* it must hide, can only lead to a breakdown in home-front confidence and the very echoes from Vietnam that the Pentagon fears the most."[15] The American Society of Newspaper Editors filed a protest with the Pentagon, asking that the formal "security review" of coverage from the field be replaced by security guidelines that journalists would follow on their own. The war ended so quickly that the Defense Department was free to treat these objections as moot and did not bother to respond substantively to them.

If a war were to last longer and U.S. casualties were higher than has been the case in recent conflicts, perhaps the public would want more information, and so opinion would swing away from support for controls on coverage. But in a short and relatively painless war with firm political leadership and public support, the Pentagon can probably get away with almost anything it wants concerning restrictions on the news media. That might not be in line with First Amendment principles, but the press and the military are well aware of the political realities surrounding these issues.

To a considerable extent technology shapes the tone and substance of the military–press relationship. Technology is at the root of many of the military's security concerns about news coverage. Real-time communication capabilities have made war a far more transparent enterprise than military officials have been accustomed to. In the Falklands war, British journalists needed Royal Navy facilities to get their stories home. By the time the Gulf war began, journalists could send their reports independently but the size of their gear limited their freedom of movement.

Change has continued. Today's journalism technology is more than an array of flashy toys. It provides its users with unprecedented independence, and that has led military planners to search for new ways to retain some control over coverage.

Journalists as Techies

Homer and Thucydides were pretty good reporters. They and others of their profession have found that the public is always eager to get news of war, be it in the form of epic poem, detailed history, or live bulletin. For Americans, the first electronic "living-room war" was World War II as brought into their homes by Edward R. Murrow, reporting from the midst of battered London. He delivered the sounds of bombs and anti-aircraft fire and even the tap-tap of Londoners' footsteps as they walked through their rubble-strewn streets. His reports were echoed a half century later as Bernard Shaw and his CNN colleagues described the U.S. air attacks on Baghdad that began the Gulf war.

The CNN team used a "four-wire" telephone, allowing them to circumvent Iraq's central phone system, which was quickly put out of commission by the air strikes. They could connect directly with Amman, Jordan, and from there, their voices traveled to the United States and the rest of the world. Twelve years later, during the most recent war against Iraq, there was no need for such devices. For the news media, it was a satellite war.

Satellites have long been used to transmit news. As early as 1963, live television coverage of John F. Kennedy's funeral was sent to Europe via the Relay satellite, but until recently the equipment for satellite broadcasting was cumbersome and expensive. Even in the early 1990s, the cost for satellite video transmission gear was about $100,000, which meant that only major news organizations could afford their own set-up. By 2003, the cost had dropped to $20,000, which let even local television stations become players. The demand for satellite time was so great during the Iraq war that Inmarsat, the firm that operates the transmission system, activated a fifth satellite to handle the traffic. The equipment is also getting smaller. During the Gulf war, a truck was needed to transport it. Soon thereafter, the gear shrank to suitcase size and weighed 60 pounds. By the start of the Iraq war, it was down to briefcase size and 15 pounds, and video transmission cost less than six dollars per minute. When the next generation of equipment becomes available, probably in 2005, it will cost about $3,000, be the size of a laptop computer, and weigh five pounds.[16]

The worldwide audience has come to expect real-time reporting, and that presents a challenge to news executives who plan coverage. Ian Ritchie, CEO of Associated Press Television News, allocated 40 cameras for coverage from Iraq and neighboring countries. His journalists in particularly remote places could use satellite telephones, some with "store and forward" capability, which involves connecting the camera to a laptop with high compression software that enhances picture quality. Because there was so much demand for live material, said Ritchie, APTN expanded its offerings from a single continuous feed to three separate channels, with almost all of the material fed live from APTN's various war zone positions.[17]

One of the most striking improvements in real-time technology was the ability to report live while on the move. CNN used a videophone connected to an enclosed antenna with a gyroscope-controlled platform that kept the antenna pointed toward the satellite regardless of the movement of the journalist's land vehicle or ship at sea. NBC's David Bloom reported live while traveling in a convoy at up to 50 miles per hour by sending his signal to an uplink truck two miles behind. The truck carried a gyroscope-aided satellite dish encased in a dome.[18]

In Iraq as elsewhere, the commitment to real-time coverage had a drawback in that it kept reporters tethered to their equipment. If correspondents are expected to provide live reports throughout the day, they cannot stray far from their transmission gear. They also found themselves cleaning out sand, recharging batteries, and otherwise nursing their equipment, using up time that could have been better spent gathering information.[19]

But the advantages offered to news gatherers by high tech tools should not be underrated. By 2003, satellite photography was more precise and more widely available than ever before. Political communications scholar Steven Livingston noted that until recently, access to these images was "limited to the national security elite holding high-level security clearances. Today, anyone with Internet access and a credit card" can order them.[20] News organizations are taking advantage of that.

Private companies such as Space Imaging employ satellites that can provide one-meter resolution pictures, which means they see objects one meter or larger. This is comparable to what an airline passenger sees from ten thousand feet. The newest commercial satellites can do even better than that. Digital-Globe's QuickBird satellite, launched in 2001, offers 60-centimeter resolution.

The satellites take a continuous stream of pictures and the field of vision is about six-miles wide. But the satellites operate in fixed orbits and so cannot "chase" a story. Also, getting photos in hand usually takes several days.

Nevertheless, these images can provide a new perspective, offering a broad visual context very different from that provided by photography on the ground. News organizations are not alone in looking at events from this viewpoint. Governments may try to use satellite photos for propaganda purposes, as Israel did in 2002 when close-ups in news coverage showed devastated Palestinian homes in Jenin on the West Bank. Satellite images, however, substantiated official Israeli reports that only a small part of the area had been targeted.

Sometimes the view from above might be misleading. Military planners who know when the satellites will be passing overhead could order tank turrets reversed to give the impression that the tanks are moving in one direction when they actually are moving in another. Also, interpretation of the satellite images is often debatable, as was seen when Secretary of State Colin Powell used satellite photos to support his case against Iraq before the United Nations Security Council. He said that the photos showed evidence of chemical and biological weapons production. Others said that they did not.[21]

During the Iraq war, *The New York Times, USA Today,* the *Los Angeles Times,* and other news organizations used satellite photos of Baghdad, Tikrit, and elsewhere in Iraq to illustrate news stories. When control of Baghdad's airport was in dispute, the news media were able to show the public a satellite's-eye view of the situation on the ground.

The news media's access to satellite imagery raises a new and possibly contentious issue in press–military relations. The military is wary about losing its monopoly on "sky-spy" technology and is also concerned about having to stand in line to get images that its own satellites might not be able to provide. During the first months of fighting in Afghanistan, the U.S. government purchased all of Space Imaging's orbit time over Afghanistan and Pakistan to ensure that the Pentagon would have exclusive access to all available satellite photos.[22]

Such tactics will not work for long. Canada, India, and other nations have commercial satellites aloft or soon will have them. Governments, private businesses, and anyone else—including terrorist groups—with a few thousand dollars can buy satellite photos. News organizations are sure to be frequent purchasers and must decide how best to use these photos, recognizing that just because something can be seen in an intriguing satellite photo does not mean that it is true.

Technological deception can also occur at ground level. The tricks that can be played with photography were displayed in the striking image that appeared on the front page of the *Los Angeles Times* and the *Hartford Courant* on March 31, 2003. A British soldier was shown warning a crowd of Iraqi civilians to take cover during the fighting near Basra. One of the Iraqi men in the foreground was holding a small child wrapped in a blanket. A fine photograph, except that it wasn't real. *Los Angeles Times* photographer Brian Walski had two shots: one with the soldier gesturing and a second with the man and child in the center of the frame. On his laptop computer, Walski used the best parts of each shot to make one image and sent it to Los Angeles. Someone at the *Courant* noticed that some people in the background appeared twice, and notified the *Los Angeles Times*. When an editor called Walski, he admitted that he had digitally altered the image and he was immediately fired.[23]

The transgression is clear. A news photograph is supposed to be an image of reality, not an improved version of reality. Digital technology makes manipulation easy to do and difficult to detect, and Walski is certainly not the only photojournalist who has tampered with images that end up reaching millions. But this was a simple call for his editors: He was wrong, and the public's trust in the news product was damaged.

The Pentagon's Plan

As war with Iraq was drawing closer, the Pentagon faced a press corps that was determined not to be muscled as it had been during the Gulf war. That determination in itself didn't much bother Defense Department officials, but it was backed up by technology that would give journalists enough freedom of movement to ensure that restricting their access to the fighting would be difficult.

So the Pentagon settled on a plan that was advertised as meeting both military and press needs. Journalists were to have "minimally restrictive access to U.S. air, ground, and naval forces through embedding," which meant that journalists would "live, work, and travel" with military units for weeks or even months. Approximately 700 slots were to be available for embedded journalists, including some from non-U.S. news organizations.

The governing philosophy, as stated by the Pentagon, was this: "Our ultimate strategic success in bringing peace and security to this region will come in our long-term commitment to supporting our democratic ideals. We need to tell the factual story—good or bad—before others seed the media with disinformation and distortions." The audience for this effort, said the Pentagon, was not only the American public but also "the public in allied countries whose opinion can affect the durability of our coalition, and publics in countries where we conduct operations, whose perceptions of us can affect the cost and duration of our involvement."[24]

The guidelines cite the importance of balancing "the need for media access with the need for operational security" and prescribe rules for news gathering. No specific communications equipment would be banned, but in "a combat/hostile environment" a unit commander could restrict transmission of information. All the embedded journalists were to accompany combat operations, and "the personal safety of correspondents is not a reason to exclude them from combat areas. . . . Gender will not be an excluding factor under any circumstance." The journalists would not be issued regular uniforms, but they would receive "NBC suits" to protect against nuclear, biological, and chemical weapons.

Certain information, such as specific numbers of troops and plans for upcoming operations, could not be released, and media representatives were reminded of "the sensitivity of using names of individual casualties or photographs," at least until next of kin were notified.

Overall, the guidelines are reasonable. The limitation on transmitting information in the midst of a combat operation is a common-sense element of responsible real-time journalism. The enemy cannot be allowed the advantage of watching the action from the perspective of the American forces. Even without a Pentagon policy on such matters, news organizations could presumably deal with these matters on their own. These are basic issues of professional responsibility, and in most—but not all—cases during the war the news media met these standards. Every so often, a journalist acted mindlessly. While doing a live shot from Iraq, Fox News correspondent Geraldo Rivera drew a map in the sand that illustrated the position of the U.S. units he was accompanying. That got Rivera booted out of Iraq (although briefly), and few journalists would quarrel with the expulsion of any colleague who endangered U.S. troops that way. During the war, 35 members of the news media, about a dozen of whom were embedded, were asked to leave or were escorted out of the war zone by U.S. authorities.[25]

Pentagon spokespersons talked about "a free press in a free society," but also had pragmatic reasons for wanting extensive news coverage. Air Force Lieutenant Colonel Larry Cox, chief of the press desk of the U.S. military command in Kuwait, cited experiences in Afghanistan in which "the Taliban and to some extent Al-Qaeda made aggressive use of propaganda. . . . In instances where there were no civilian casualties or collateral damage they had the ability to invent them, and no one would know the difference until after the press coverage resonated in the world community. . . . We needed to have the maximum possible access of a free press operating on the battlefield, not controlled by the U.S. or the coalition, but in position to do third-person objective reporting that we knew would . . . illuminate lies and exaggerations."[26] Some observers said that another purpose of the Pentagon's openness may have been to "weaponize" news coverage in the sense that positive, close-up reports about American military power might demoralize the enemy and speed up surrender by Iraqi forces.[27]

The Pentagon arrived at its number of roughly 700 slots for embedded journalists based on logistics, said Cox. There were just so many journalists, he said, "that a unit could absorb before [they] were a hindrance." Also, news organizations had just so many journalists to deploy.

Once the news media had accepted the embedding plan, the Defense Department provided plenty of support. The Pentagon offered training, transportation, food, shelter, and of course protection, all worth thousands of dollars and all paid for by taxpayers, not news organizations. News executives generally claimed necessity as the reason for this coziness, but even when assistance is accepted its fair market value can be determined and news organizations could reimburse the government for value received. Failing to do so is the kind of thing that undermines the public's perception of journalists' independence.[28]

In the natural order of things, the press tells what it knows and the military embraces secrecy. Those fundamental traits ensure a persistent tension in press–military relations. Because the Pentagon's ground rules were seen by some journalists as constraining the independence of embedded reporters, news organizations devised a two-tier coverage plan. While some correspondents would be embedded, others would be "unilaterals" who would operate on their own.

Pentagon officials made clear that the embedded journalists would receive better treatment than the approximately 1,800 unilaterals. When Kuwait blocked some unembedded reporters from entering Iraq, Pentagon spokesperson Bryan Whitman said: "We are going to control the battle space. Reporters that are not embedded are going to be treated like any other civilian, approached with a certain amount of caution. For many journalists, proving their identity can sometimes be problematic." Whitman raised the possibility that the Iraqis might have "individuals pose as journalists."[29]

Michael Massing, who was in the war zone to monitor the fighting for the Committee to Protect Journalists, found the military unconcerned about the safety of unilaterals. He said that "the U.S. military believed that only reporters who were officially embedded had the right to protection. Everyone else was at risk—and expendable."[30]

The Pentagon had a product to sell and officials saw the embedding plan as the best way to do it. As Larry Cox said, "One of the beauties of being embedded is learning about the personality of the unit, about the color and the depth, the substance that you don't usually get if you're not associated with a unit in that way."[31]

That may have been true, but journalists had to decide if that access was really an improvement over detached observation.

Embedded Journalists at Work

Bomb it and they will come, especially to cable TV. With their intensive coverage of the Iraq war, cable news channels dramatically increased their audience. Nielsen Media Research found that the number of average daily U.S. viewers of MSNBC and CNN increased more than 300 percent and the Fox News audience rose 288 percent. Fox News was the most-viewed cable channel, with an average of 3.3 million viewers per day. Among the broadcast networks, NBC led with more than 11.3 million viewers per day.[32]

This coverage didn't come cheap. Major networks were estimated to be spending a million dollars a day on covering the war. Ad revenues, meanwhile, dropped precipitously during the war's first week as networks cancelled spots to provide uninterrupted coverage and advertisers pulled ads from the slots that were available because they feared that their messages would seem inappropriate. During the first week, television ad revenues were down $77 million compared to the previous year.

But over time, war makes money for many, including the television industry. By the war's second week, when slots for ads opened up and viewership remained high, the ads poured in, running more than $70 million ahead of the previous year.[33] Prime-time entertainment shows that were preempted by the war could be rescheduled later in the year, saving costs at that time. The prime-time TV magazine shows could rely on war coverage for their content. And while news organizations were spending a lot to cover the war, they had little need to cover much else, and so could cut those costs. Former NBC News president Lawrence Grossman noted that covering war and other disasters "is the most efficient way a news division can spend its money, because so much of what the money is spent on gets on the air."[34]

Although financial concerns might not seem compatible with the noble work of delivering the news in a time of crisis, this is a big part of the reality of the news business. The quantity and quality of war reporting are linked to news organizations' profit margins.

Meanwhile, far removed from the corporate accountants, embedded reporters were delivering a remarkable product that often provided the audience with vivid depictions of the fighting:

- Ron Martz of the *Atlanta Journal-Constitution* telling how his photographer held an intravenous drip bag over a wounded Iraqi civilian and later describing how two American soldiers at his side were wounded as they entered Baghdad.
- Walter Rodgers of CNN presenting "live pictures of the Seventh Cavalry racing across the desert in southern Iraq."
- Dexter Filkins of *The New York Times* telling of a sergeant's reaction when his unit shot a woman standing next to an Iraqi soldier: "I'm sorry, but the chick was in the way."

A content analysis conducted by the Project for Excellence in Journalism found that American television stories from embedded reporters during the first week of the war featured "all the virtues and vices of reporting only what you see." The study found the following characteristics:

- 94 percent of the stories were primarily factual rather than interpretive.
- 60 percent of the reports were live and unedited.
- In 80 percent of the stories, viewers heard only from the reporters, not from soldiers or other sources.
- 47 percent of the coverage described battles or their results.
- The reports avoided graphic material; not one of the stories in the study showed pictures of people being hit by weapons fire.

The report found that the war as presented by embedded television journalists was like "reality itself—confusing, incomplete, sometimes numbing, sometimes intense, and not given to simple story lines."[35]

Reporters reflecting on their embedded coverage described the conflicting loyalties and one-dimensional information that they had to work with. Sometimes the decision about what to report was made less difficult by the embedding agreement that reporters and editors had signed stating that information would not be disclosed if it "may be of operational value to an adversary." *The Washington Post*'s Steve Vogel, embedded in an airborne brigade in northern Iraq, was asked by the commander not to report that tank reinforcements had not arrived. The reason: If the Iraqis knew that the relatively small U.S. force lacked armor, they might attack. Vogel later wrote: "I wanted no part of a deception campaign. But I also did not want to expose a vulnerability that could get paratroopers killed. I reported what the brigade had, and not what it didn't have. . . . Until they began arriving April 8, there was no mention of tanks."

Sometimes the issues were less clear. Mary Beth Sheridan, also of *The Post,* was embedded with a U.S. helicopter brigade, and the pilots frequently told her, "We don't shoot civilians." She wrote that "the soldiers had mantras that were so widely held they were a kind of group-think. 'We don't kill civilians.' 'We're here to help.' I don't know how these ideas were transmitted. But on the hermetic military bases, islands with little news and few outsiders, they went unchallenged. Without realizing it, you could get taken in by their narratives, to think that war wasn't messy." After the war had been under way for more than two weeks, wrote Sheridan, she finally met her first Iraqis—a family that had been in a taxi that was fired on by an American tank. Two adults and two children wounded, one adult dead. That was the messy reality that Sheridan had not seen before.

The Post's Lyndsey Layton lived for a month on the aircraft carrier *USS Abraham Lincoln,* spending most of her time with a squadron of fighter pilots, often just hanging out with them in their ready room so she could, she said, "become a familiar face." That led the pilots to treat her more as a buddy than a reporter, and one morning she found one of the pilots exuberant about the success of his bombing mission the night before. That raised questions for Layton. She later wrote: "Would he want the world to know he was buzzed on bombings? I doubt it. . . . Was it fair to shine a spotlight on those unguarded moments? On the other hand, isn't that why I was there? To get as close to the 'truth' as possible?" In her final story from the ship, wrote Layton, she "included a reference to this episode but I handled it gingerly and didn't completely report what I saw. I'm still not sure whether I made the right choice then, or now."[36]

Gordon Dillow of the *Orange County Register,* who had been an Army sergeant in the Vietnam war, said that being embedded affected his perspective, making it similar to that of the troops. "Isolated from everyone else," he said, "you start to see your small corner of the world the same way they do." That affected the content of his stories. "I didn't hide anything," Dillow later wrote. "For example, when some of my marines fired up a civilian vehicle that was bearing down on them, killing three unarmed Iraqi men, I reported it—but I didn't lead my story with it, and I was careful to put it in the context of scared young men trying to protect themselves. Or when my marines laughed about how .50-caliber machine gun bullets had torn apart an Iraqi soldier's body, I wrote about it, but in the context of sweet-faced, all-American boys hardened by a war that wasn't of their making. And so on. The point wasn't that I wasn't reporting the truth; the point was that I was reporting the marine grunt truth—which had also become my truth."[37]

"My marines." "My truth." Dillow's choice of words shows how sometimes the line between journalist and soldier became blurred, both as a matter of self-image and in story content. Scott Bernard Nelson of *The Boston Globe* wrote that as the war wound down "a funny thing happened on the way home from Iraq this week: I found myself scoffing at the rear-echelon soldiers for how little they knew about the war. About the real war, the one I had experienced, with enemy AK–47 rounds buzzing over your head and the smell of burning flesh and metal filling your nose." Nelson, who before the war worked for *The Globe*'s business section, had the grace to include some self-mockery as he described his transformation into warrior-reporter, but there was substance to his claim of being hardened by combat. He reported that he "did more than just empathize with soldiers. I helped them in battle." When the Marine unit in which he was embedded came under fire, Nelson saw muzzle flashes coming from a nearby building. He yelled to the gunner of the vehicle he was in, leaned out the window, and pointed out the location. The gunner then blasted the spot with his .50-caliber machine gun, killing the sniper.

Reflecting on his experiences, Nelson said: "Whether I acted out of self-preservation that day or because of an affinity with the soldiers I was covering hardly matters. The question is whether the coverage I provided during the war was tainted as a result. I'd like to believe it wasn't." He concluded by noting his hope that his was just one of many diverse voices that *The Globe* presented in a balanced way.[38]

Other reporters also found that the realities of battle can supersede conventional standards of journalism. Gordon Dillow reported that in the midst of a prolonged firefight in Baghdad, a marine gave him a hand grenade to throw if the enemy started to overwhelm the U.S. unit. Dillow wrote: "I know that my having it violated written and unwritten rules. Still, it felt comforting in my hand. (I never had occasion to throw it.)"[39]

Nelson and Dillow had gotten inside the story. In such cases, wrote *The Washington Post*'s Peter Baker, the embedding system provided "fly-on-the-wall access journalists always seek and so rarely obtain." As for the relationship between journalist and source, Baker wrote that "instead of automatic mistrust, now perhaps we understand each other just a bit better. Instead of assuming the worst, now perhaps we'll give each other a chance. A generation of reporters and soldiers who grew up deeply cynical about each other after Vietnam has built more of a normal dynamic."[40]

Journalists who operated outside the embedding system were not convinced that their embedded colleagues' closeness to sources led to better stories. Edward Gargan of *Newsday* said: "It's like being in a cocoon. You really have an umbilical cord to your unit. That's not the kind of reporting I want to do."[41] ABC's John Donvan raised the issue of Iraqis viewing embedded journalists as part of the military units they accompanied. "The Iraqis," said Donvan, "are now seeing journalists dressed up in camouflage and riding with the troops, and it's hard for them to make the distinction we like to make—that we are civilians, defenseless, fair brokers, and we want to hear both sides of the story." Donvan added that when trying to do stories about Iraqi civilians, "the whole hearts-and-minds thing cannot be done from inside an embed."[42]

The basic trade-off, then, is between access and independence. The best resolution for news organizations is to use both embedded and unilateral journalists. Peter Baker said that "the embed-versus-unilateral debate is a false construct. It presumes that one is better than the other. In fact, each has its strengths and weaknesses, and the reader is best served by having both. The coverage was richer because embedded reporters were able to go places with the military and see things with their own eyes, and because unilateral reporters were able to talk to Iraqis and search out stories their colleagues tied to fast-moving military units invariably would miss."[43]

British news organizations also had journalists embedded with the Iraqi invasion force, and a study of their coverage by a research team at the Cardiff School of Journalism found that most of the embedded reporters "made efforts to protect their objectivity and, on key issues, were able to do so." The embeds, according to the study, "revealed, on many occasions, how unreliable the information from military briefings can be" and they "were often able to correct misleading claims by military sources." The study also found, however, that the television coverage provided by embedded journalists was "full of action, but without the grisly consequences." Overall, the British news coverage was also found to have lacked skepticism about claims by the U.S. and British governments concerning weapons of mass destruction and was susceptible to spin from officials of the two governments.[44]

Another issue was personal safety. Between March 22 and late April, the period of most intensive fighting, 14 journalists of the roughly 1,500 in the war zone were killed, which was a far greater casualty rate than that of the U.S. troops. (More would die later, as the war dragged on.) Six of them were killed by Iraqi fire, four by friendly fire, three in accidents, and one due to a medical condition that probably was aggravated by the working conditions.[45] Battlefield dangers led some unembedded journalists to hook up with military units, and they were particularly appreciative of the relative safety this provided. ABC's Donvan said: "These people have been so nice to us, I could see how it'd be hard to write critically if they made a mistake. I don't think I could do it. They're my protectors."[46]

After U.S. forces seized Baghdad, embedded journalists began peeling off from their assigned units and operating on their own. By late April, fewer than 200 journalists remained embedded. By September 2003, the number was down to about two dozen.[47]

The American public also made some midstream changes as it watched the war. A substantial number of television viewers switched allegiance from the major broadcast networks to cable news. After the first few days of the war, CBS and ABC together lost about two million viewers, a combined 10 percent. NBC, which alone of the big three has a cable news operation, saw a slight increase in audience. Solid conclusions about this are difficult to reach because some of the numbers are hard to read; shows like ABC's *Nightline* added viewers while the network's total news audience diminished. But overall this was a remarkable shift in viewing preference. In the past, a major news event such as a war or the 9/11 attacks brought larger audiences to network news. Never before while such major stories unfolded have the big three evening newscasts seen their viewership decline.

A number of factors contributed to this. Because of the time difference, much of the fighting in Iraq happened during late night in the United States and the cable news channels could deliver live coverage whenever events warranted, which was less feasible for the broadcast networks. Also, the role of the network anchor changed. Andrew Heyward, president of CBS News, said that "after 9/11, viewers looked to the network anchormen to help knit the fabric of the nation back together." But in contrast, he said, Iraq "was a reporters' war, not an anchor war," and that let the cable channels compete more effectively.[48]

The news audience was looking for real-time, dramatic stories, and the large press corps in Iraq was providing lots of them. There was no doubt about the quantity of news available, but questions remained about the quality of the coverage.

What Was Missing?

As the embedded journalists roared through the desert with their units, they produced gripping pictures and descriptions of the war. Their news organizations were happy because they knew that this was the kind of material that holds an audience, especially television viewers. It was "reality" programming, something like "COPS Goes to Baghdad." The Pentagon was happy because the competence and bravery of the U.S. armed forces were on display, showing

Americans what their tax dollars were buying, and putting the rest of the world on notice that the U.S. military was unstoppable. The American public was happy, too, proudly watching a helluva show.

But how good was the journalism, in the sense of not just telling the audience what was happening in a particular place at a particular moment, but also helping people understand matters such as why the war was being fought and what its ramifications were? Jack Fuller, president of Tribune Publishing Co., said that the television coverage was "utterly riveting," but that "it also demonstrated that there is a difference between seeing and understanding."[49]

One frequently heard description of the coverage was that it was like seeing the war through 600 soda straws—an array of tightly focused yet very narrow viewpoints. Even what was seen through all those straws might not have been representative of the larger picture. Secretary of Defense Donald Rumsfeld referred to the "slices of war" that the news media were presenting and cautioned against reading too much into any particular slice. News executives who took time to reflect on the nature of their product also were wary of overselling. Erik Sorenson, president of MSNBC, wondered, "Who knows how much the embedded reporters saw. Did we see 8 percent of what happened? Did we see 4 percent of what happened? It's arguable they didn't see a double-digit percentage of what happened."[50]

This was not the fault of the journalists on the ground. They were doing precisely what they were supposed to do: tell what was happening at their tiny piece of the war. The problem was in the upper ranks of news organizations, where producers and editors became so infatuated with the gritty drama being delivered by the embeds that they overused that to the exclusion of less gripping but more important stories. Combat is only part of war; the rest of it is politics, diplomacy, economics, and other dry sciences. But much of the news coverage of the war, especially television's version, did not reflect this.

Also lost in much of the coverage was any connection between what was happening and the reasons for the war. Occasionally a cache of Iraqi chemical warfare protective suits and antidote syringes would be found and cited as evidence that Saddam Hussein's forces might use "weapons of mass destruction" against American troops. But during the advance on Baghdad no such weapons were found. Reporters were eager to speculate, however, about even vaguely suspicious items such as metal drums with unknown contents, which on one occasion turned out to contain nothing more sinister than pesticide. As the U.S. invasion pushed ahead, substantive evidence was found of Saddam Hussein's brutality, such as mass graves and torture chambers. But rarely was the public reminded by news stories that getting rid of Hussein for those reasons, although a good idea, had not been cited by the U.S. government as the principal reason for going to war.

On such matters the Bush administration received a free ride because the mainstream news media apparently decided that it would be inappropriate—perhaps even unpatriotic—to challenge the government while the war was under way. But as Michael Dobbs of *The Washington Post* pointed out, one of the reporter's jobs is to be "an antidote to the propaganda."[51] That requires not being complicit through passivity.

The coverage also made some narrow and premature judgments, ascribing greater significance to particular events than was due. One embedded reporter's unit encountered Iraqi soldiers fleeing Basra, ready to surrender. The report, which ran at the top of a network newscast, gave the impression of widespread, spontaneous surrender by the Iraqis. That was simply not true. Resistance continued and Basra was not yet controlled by U.S. or British forces.[52] One reporter's snapshot of events became transformed into a definitive panoramic view. That reflects sloppy judgment by editors in too much of a hurry.

The toppling of the big statue of Saddam Hussein in Baghdad was portrayed in much coverage as symbolizing the end of the regime and therefore of the war. The air of celebration and the friendly interaction between Baghdad residents and U.S. soldiers at that particular time and place reinforced this impression. But although the eventual outcome was not in doubt, the war was not over. Pentagon spokesperson Bryan Whitman recalled that "at that point it was not representative of Baghdad, and certainly not of all Iraq. We were having to really remind people that Baghdad was still an unsafe environment and there was still much fighting going on."[53]

Some news organizations took note of the need to do more than simply present a stack of individual stories. Each might be accurate in terms of the facts it presented, but taken together they would be disjointed and not add up to a coherent whole. To address this, ABC's *Nightline* presented a segment each night titled "The Big Picture," which assessed the overall state of the war.

Emphasis on real-time reporting was not unique to this war; it has been transforming the news business for some time. Particular attention is paid to fast pacing within individual stories and in newscasts. CNN's Christiane Amanpour said that "our network has gotten away from taped packages; they think 'live' brings more spontaneity. 'Keep it moving, keep it moving' is what they tell us."[54]

That changes the texture of news. The live story is often incomplete and breathlessness takes the place of good writing. Doing a story after the key events are over does not mean the reporting will be stale. Instead the telling may benefit from reflection and having time to find the words and pictures that best convey the meaning of those events.

Placing such a high priority on speed will inevitably lead to errors. By the end of the first week of the war, *Editor and Publisher* had published a list of "fifteen stories [the news media] have already bungled."[55] Most dealt with whether certain cities had been taken or whether Iraqi resistance was collapsing. In most cases, the news organizations at fault corrected their mistakes, but that does not excuse the fact that bad information about significant events was being delivered to the public.

Consider the initial coverage of the March 22 grenade attack at Camp Pennsylvania in Kuwait. Fox News reporter Doug Luzader in Kuwait cited the Associated Press and Reuters when he told anchor Tony Snow that "apparently one or more terrorists infiltrated the perimeter of this camp." Luzader said that the attack showed that "not only are these camps somewhat vulnerable, but also the fact that they may have terrorist operatives, perhaps even coming in from Iraq over the border." After this speculation, he added, "Now, there's no indication that's what happened here." Fox then got Sky News reporter Stuart Ramsey on the phone from the camp where the attack occurred. He provided this account: "It seems that two Kuwaiti or Arab nationals entered the headquarters tent in Camp Pennsylvania. . . . It seems, and this is not confirmed at the moment, that these two foreign nationals were possibly on the translating staff or working at the headquarters. . . . These two men were apparently wearing desert camouflage gear." Anchor Snow then summed up the story of the attack as "presumably by a couple of men who were serving as translators."[56]

That sounds like good, prompt reporting, using a number of sources. The problem is that the only accurate item in the story is that there was an attack. Nobody came across the border, with or without camouflage uniforms, and no Kuwaiti or other Arab translator was involved. An American soldier was arrested and charged with two counts of murder and seventeen counts of attempted murder.

Fox News was not the only news organization that delivered the inaccurate story. *The Washington Post* and others got it wrong at first. They all eventually reported the real story, but why were they so eager to push speculation as news? Why is there an aversion to saying, "We don't yet know what really happened"? The answer is found in the new culture of journalism that reveres being first above all else. Wanting to beat the competition is nothing new, but there should be systemic safeguards built on basics such as verifying facts and corroborating sources. Instead, the old wire service adage "Get it first but first get it right" has been pushed aside, consigned to the basement with the linotype machines and other dust-covered relics.

Plenty of other mistakes were made. For instance, the thousand-vehicle convoy of elite Iraqi troops that was reported to be on its way to counterattack American troops simply did not exist. On stories like that, finding the truth was not always easy. U.S. military spokespersons sometimes were unable to comment on queries from reporters because no one had anticipated just how fast information would travel. As good as U.S. military communications were, the news media's were even better. Paul Slavin, executive producer of ABC's *World News Tonight* said that journalists on the front lines would call their colleagues with information, and when these reporters went to Central Command for confirmation officials there would say, "We don't know anything about it."[57] The reporters then asked why the military was trying to cover up the truth. Brigadier General Vincent Brooks, who conducted many of the Central Command briefings, explained: "That becomes the perception. It's just a very interesting dynamic."[58]

Speed may be a principal cause of mistakes, but another factor is also sometimes at work—a schadenfreude that can distort journalists' appraisals of government policy. After a few days of their fast-moving advance into Iraq, American forces paused to strengthen supply lines and deal with pockets of Iraqi resistance. Instantly, news coverage was filled with loud questions about the wisdom of the overall war strategy. Defense Secretary Rumsfeld and General Tommy Franks were targets of nasty potshots from journalists and their hired analysts, mostly former military officers. The dreaded word "quagmire" emerged and other analogies to Vietnam were dragged out of the media's closet.

Press criticism of policy is fine, but it should be based on fact. The idea that the pause in the thrust toward Baghdad was the beginning of a prolonged Vietnam-like disaster for U.S. forces had no basis in military reality. As the next stage of the war proved, the American invasion strategy was a success—objectives were achieved quickly and at relatively low cost in U.S. casualties. The pause had been simply . . . a pause.

Thomas Kunkel, dean of the University of Maryland's college of journalism, wrote that "the media showed an almost palpable need to rush to judgment. . . . How can you intelligently discuss military strategy four days into a war as complex as the invasion of Iraq? It seems to me this was another of those areas where the people, in their collective wisdom, were way ahead of the press. They implicitly understand that wars take time. While reporters were raising hell about tactics, every public survey demonstrated the American people were willing to let things play out before they made up their minds. We were the impatient ones."[59]

Impatient and underinformed. In journalists such a combination is certain to produce flawed coverage. Demand for speed keeps coverage shallow and quasi-political negativity shapes perspective on events.

These are not esoteric concerns. They diminish truth, the essential element of news. During the Iraq war, another aspect of media behavior further affected truthfulness: sanitized coverage that distorts the reality of combat.

The Project for Excellence in Journalism's study of the first week of the war found that even with embedded journalists on the front lines, there were limits to the realism of the coverage. Twenty-one percent of the embedded television journalists' stories showed combat action—weapons being fired—and in half of those stories, viewers saw the firing hit buildings or vehicles. But none of the stories showed Americans or Iraqis being killed or injured by weapons fire. The study reported that still photos published in newspapers were often more graphic than the television images, and video on non-American television was also more graphic than that which U.S. channels provided.[60]

Real-time technology adds another layer of complexity to questions about how much to show. With television now able to cover combat live, news executives must consider the prospect of people being killed or wounded on the air—"live death." If the fighting is being covered live, showing casualties as they occur might be unavoidable. Marcy McGinnis, CBS's senior vice president of news coverage, noted that "you could be filming a firefight live and somebody falls in front of us. You'd have to make the decision if you'd show that again on tape or not. We would not want to be inappropriate or tasteless. That being said, we're covering a war, so we're not going to never show the dead or never show the wounded."[61]

Print media faced similar decisions. During the Iraq war, *Army Times* (despite its name, a privately owned newspaper) ran a picture of a seriously wounded American as fellow soldiers carried him away from the fighting. Anyone who knew him would have recognized him in the photo. He died the next day. In a letter to the paper, Secretary of the Army Thomas White and four other officials criticized the *Army Times*'s "callous disregard for basic standards of decency and the emotions of the family and loved ones of this brave soldier in their time of grief. . . ."

In an editorial, the paper said that deciding to run the picture was "painful," but the photo "bespeaks a fundamental truth. . . . He was a real man, with a real face, who died on the battlefield. . . . This picture helps ensure no one forgets that."[62]

Policies about covering civilian casualties are also controversial. "Collateral damage" is the loathsome shorthand for innocents who are trapped in the wrong place at the wrong time. When shown, their bodies are reminders of how horrible war really is, but the news media frequently pretend they do not exist, or else reduce their deaths or maiming to numbers on a tidy scorecard—a box with casualty figures stuck somewhere in the newspaper or flashed briefly on the television screen.

In coverage of the bombing of Baghdad, video was presented with a technology-enhanced clinical detachment. Cameras were positioned on rooftops, links to satellites established, and then . . . Let the fireworks begin! *Newsweek*'s Christopher Dickey said that this kind of coverage "contributes to this notion of war as video game and strips the war of its humanity."[63]

The images were spectacular, to be sure; modern war as a son et lumiere extravaganza. But forgotten in the news coverage of the pyrotechnics were the people living in the city that was being bombed. The primary targets may have been Saddam Hussein and his palace guard, but they were not the primary victims. "Shock and awe" is a clever turn of phrase, but it has dreadful meaning for those who get caught up in it.

Journalists know they should report about the fate of civilians caught in a war zone. John Walcott, Washington bureau chief for Knight Ridder Newspapers, said that reporters are encouraged to cover all aspects of the fighting. "It is our responsibility," he said, "to show the face of war—no matter what it looks like." But *Newsday* editor Anthony Marro admits, "We pay more attention to American deaths. It is easier to report on people we know. . . ." Tim Connolly, international editor at *The Dallas Morning News,* said that when the paper ran a page one photo of a Baghdad market bombing, he found "some people view coverage of the victims of war as being antiwar. . . ."[64]

When civilian casualties *are* reported, how graphic should the coverage be? G. Jefferson Price, editor of *The Baltimore Sun*'s "Perspective" section, faced this question concerning a photo of an Iraqi man carrying an injured young girl in his arms. The issue was whether to crop the photo so as not to show the bloody stump where the lower part of the girl's leg had been blown away, apparently by a coalition bomb at Basra. Editors are cautious about such things because, said Price, "people look at their newspapers in the morning and we don't want to upset our readers' breakfasts." Price said, however, that he believed that "we should show the whole ghastly image to let people know how inhuman and horrifying war is. I also believe we should show images of the dead, whether they are Iraqis or coalition troops. Death is what war is about. People who believe that war is good and necessary need to see these images. People on both sides."

Nevertheless, caution won out and when *The Sun* ran the picture of the girl, they did so without showing the wound. Price then received a message from an angry reader who had seen the whole picture in a Canadian newspaper and wondered why *The Sun* was trying to censor the truth. Price explained that "we do that to be unoffensive. But it does hide the full truth and probably we shouldn't do that."[65]

Another facet of the coverage of civilian casualties was determining whom to blame. The Pentagon was understandably sensitive about allegations— delivered to the entire world by the international press corps—concerning reckless targeting of air strikes and artillery fire. U.S. military spokespersons said that some of the civilian casualties had been inflicted by Iraqi weapons. For reporters at the scene, dealing with such charges and countercharges was difficult. The *San Francisco Chronicle*'s Robert Collier, who reported from Baghdad, said, "Although it was rarely possible to completely discount U.S. claims that the explosions that killed and maimed innocent Baghdad residents were caused by Iraqi anti-aircraft fire or had been set off deliberately to draw world sympathy, the circumstantial evidence pointed overwhelmingly toward U.S. culpability."[66] On the other hand, National Public Radio's Baghdad correspondent, Anne Garrels, found that damage to a house where civilians had been killed was "not consistent with a missile or an American bomb." A former Iraqi military officer who examined a piece of the ordnance that hit the house told Garrels that it was from an Iraqi antiaircraft gun. Garrels also interviewed a doctor (who was also a military officer) at Baghdad Neurological Hospital who said the fragments he removed from victims of an explosion at a city market were from errant Iraqi surface-to-air missiles and antiaircraft chaff.[67]

Craig Nelson, who reported from Baghdad for Cox Newspapers, said that determining the number of dead and wounded civilians was often difficult because it was hard to get to the scene before they had been buried or taken to hospitals. He noted that "under wartime conditions—when both sides have a stake in depicting their cause as just and their efforts as righteous, when non-governmental organizations and journalists have little or no timely access to the front and to bomb sites—there's no 'definitive report.' . . . Until Iraqi civilians can be interviewed without duress to determine the exact circumstances of the attacks and determine what, if any, military purposes certain buildings were being put to, there will be no 'definitive' reports."[68]

Definitive or not, the coverage found a receptive audience in the United States. During the first days of the war (March 20–22), 80 percent of respondents in a Pew Research Center poll rated the coverage as excellent or good, and by the first week in April that number was still holding at 74 percent. Twenty-three percent said the coverage was too critical of how the United States and its allies were conducting the war, while 9 percent thought the coverage was not critical enough. The polling done April 2–7 found that 39 percent of respondents said the war was getting too much coverage, at the expense of domestic issues, particularly those related to the economy. Forty percent said that anti-war sentiment was receiving too much coverage, while 28 percent said that Iraqi civilian casualties were getting too little attention. Sixty-nine percent of the respondents said they preferred neutral coverage, while 23 percent said they wanted pro-American reporting.[69]

———

Public approval can be an insidious thing, deterring the professional introspection essential to improving coverage. Embedded reporting won much applause, but journalists should consider the quality of the overall news product—breadth, depth, and sophistication—before resting on their laurels and committing to relying on the same coverage patterns for the next war.

Among the specific issues needing examination are the ties between the news media and the Pentagon. Even in wartime, that relationship should include a healthy dose of skepticism and wariness.

CHAPTER FOUR

On The Team?:

The Press and the Pentagon

From the Pentagon's perspective, embedding was the signature innovation in press–military relations during the Iraq war, and it will almost certainly carry over into the next conflict. Underlying the embedding process were larger issues:

- During wartime, what happens to reporters' adversarial approach? Should it be set aside, or at least turned down a few notches?
- To what extent do the responsibilities of citizenship supersede those of journalism?

The official line from Pentagon officials was that embedding was a success, showing American forces performing well and giving the American public a good sense of battlefield rigors. But there were complaints as well from the administration, centering on how reports from embedded journalists influenced the overall coverage. After the war had been under way for slightly more than a week, Defense Secretary Donald Rumsfeld said: "We have seen mood swings in the media from highs to lows and back again, sometimes in a single 24-hour period. For some . . . the massive volume of television—and it is massive—and breathless reports can seem to be somewhat disorienting."[1]

Those reports kept Pentagon spokespersons scrambling. Assistant Secretary of Defense Victoria Clarke noted that a flurry of stories had reported American forces advancing on Baghdad were running out of food. "One reporter," she said, "may have heard one person say, 'I haven't gotten my second or third MRE [meal ready to eat] of the day yet' and that gets back, and it gets back very quickly. . . . Because of the volume and velocity of the reporting coming directly from the field, we all have to take a breath sometimes and say, 'Okay, is this representative of a broader picture, or is this an individual incident?'"[2]

Several weeks later, Rumsfeld said coverage of the war was sometimes "inaccurate" and "conflicting," but he praised the embedded reporting, noting that "the American people . . . could see accurate presentations . . . of what the men and women in uniform were doing." The chairman of the Joint Chiefs, Air Force General Richard Myers, added his approval, observing that the embedding system "may over time get us away from some of the cynicism that has developed" in terms of how the news media view the military.[3] And after the war, Clarke reflected that "good news and bad news, transparency works. The good news gets out; the bad news gets dealt with quicker. That's a good thing."[4]

Generals as TV Stars

The reports from the embedded journalists looked particularly good to Pentagon officials when compared to some of the analysis being provided on television by retired military officers. This kind of commentary was nothing new. The networks and cable channels, perhaps sensitive to being perceived as soft on military matters, had over the years recruited telegenic military experts to discuss strategy, tactics, and weapons. Military analysts established an on-air presence during the 1991 Gulf war and had been used in conflicts since then. For the Iraq war, the commentators' numbers increased and they had plenty of work. Between March 20 and April 21, they made 214 appearances on the six principal television venues—the big three broadcast networks and CNN, Fox News, and MSNBC. CNN used the analysts the most, 55 times, and ABC the fewest, 29 times.[5]

The experts quickly came under fire from critics. David Zurawik of *The Baltimore Sun* dubbed them the "windbags of war" and said that their presence had "more to do with symbolism and staging than it does with insight." Zurawik said much of the commentary was superficial, but noted that sometimes the analysts were hampered by being squeezed into formats such as MSNBC's "military minute" that reduced important matters to "show-business triviality."[6]

More pointed criticism came from the Pentagon. Secretary Rumsfeld blasted "retired military officers" whose approach, he said, was to headline their analyses "'Henny Penny: The Sky Is Falling,' 'It's Just Terrible,' and 'Isn't It Awful?'"[7] Joint Chiefs chairman Myers said, "I think for some retired military to opine as aggressively as some have done is not helpful." After meeting with Rumsfeld and Myers, Senate Armed Services Committee chairman John Warner said that the former officers "should follow the tradition of presidents, the commanders in chief. You do not see former presidents criticizing a sitting president during war."[8] Victoria Clarke later said of the analysts: "I've always thought their role was inflated. And because of the volume and velocity of real news coming back from the war, the role of the analysts and commentators was far less."[9]

The main theme of the analysts' criticism of the war plan was that too few troops had been sent to Iraq to guarantee prompt success. Among the analysts, the toughest critic was General Barry McCaffrey, who had been commander in chief of the U.S. Armed Forces Southern Command during the Gulf war and later served as anti-drug czar in the Clinton administration. He was unrepentant in his response to Rumsfeld, stating: "This war is too important to be left to the secretary alone. At the end of the day I think they ought to value my public opinion."[10]

Some of McCaffrey's colleagues, however, decided to retreat. Former NATO commander General Wesley Clark, who had expressed his concern that not enough troops had been deployed, said that "It's very unfair and difficult for anyone to criticize a war plan without having been involved in the planning process." (When Clark became a presidential candidate later in 2003, he renewed his criticism of the war plan.) Retired Air Force Major General Burton Moore stood up for the right of former officers to comment, but admitted that morale in the field could be affected, saying, "If troops on the ground keep hearing from people who keep saying, 'We don't have enough forces, we don't have enough forces,' you can't blame some of the soldiers if they start believing it."[11]

Television news executives mostly defended their use of the analysts. Marcy McGinnis of CBS said that it was important for the public to hear more than just the Pentagon's pronouncements, and added that "generals in studios are not going to turn the tide of war or make our soldiers get into depths of depression." ABC's Amy Entelis said that the military consultants' job was to analyze, not criticize, although "analysis can be critical." But, she added, "the point is not to throw stones. It's to have a better understanding of what is happening." Erik Sorenson of MSNBC was bit more cautious, stating that "this is not a primary in New Hampshire. This is life and death, with people in harm's way, and we have to be careful" not to "undermine confidence and compromise lives on the battlefield."[12]

Although these experts were showcased in the networks' war coverage, issues related to the integrity of their quasi-journalistic product received little attention. Where were they getting the information that they used in their analyses? What relationships existed between the commentators and the Pentagon, particularly with Rumsfeld, who had clashed with some of the armed services' top brass? Were they on the payrolls of defense contractors or did they have other ties that might produce conflicts of interest? These and other such matters could have affected the content of their pronouncements and so should have been disclosed.[13]

Furthermore, these experts are hired to provide context for the bare-bones information provided by the Pentagon and the on-scene reports coming in from reporters, but do they really know what they're talking about? Certainly, they have their own experiences to use as the foundation for their opinions, but now they are outsiders and get much of their information from an old-boy network of sources who almost always remain anonymous and who may have their own agendas and conflicts of interest. Information that the commentators gather is not part of the regular reporting process and is not subject to all the checks that are used to verify news items. The experts' credentials provide nice window dressing for news organizations, but if these retired officers are to truly enhance the news product, they might concentrate on offering more fact-based historical perspective and less speculation about what might happen next.

The public found the retired officers' commentary less than enthralling. A Pew Research Center survey reported that 36 percent of respondents said they were hearing too much from the analysts, 11 percent too little, and 48 percent the right amount. The survey found that this was one topic on which supporters and opponents of the war agreed.[14]

The Pentagon's Spin

The generally pessimistic tone of the retired officers' commentary proved incorrect as U.S. forces quickly moved into Baghdad. Vice President Dick Cheney brushed off the opinions of "generals embedded in television studios," and the administration relentlessly pushed its bullish appraisals of the fighting. Many in the news media bought into this, entranced by the Pentagon's promotion of "shock and awe" as the characterization of U.S. strategy. Terence Smith, media correspondent for the PBS program *The NewsHour with Jim Lehrer,* said that the news media had consistently accepted the administration's forecasts about matters such as the strength of Iraq's Republican Guard and the prospect of a fierce defense of Baghdad while not adequately probing events such as the failed assault by a wave of Apache helicopters that resulted in one copter being shot down and its crew captured, and the rest badly damaged by ground fire.[15]

Smith also argued that even if journalists believed that the invasion plan was properly designed, "it is totally legitimate to keep asking those questions about that plan and whether or not it was flawed in this respect or in that respect. That's the duty, the obligation of news organizations in a situation like that."[16] Some media critics saw a pattern in news organizations' credulity. John MacArthur, publisher of *Harper's Magazine,* wrote that the government's public relations effort, which had built support for going to war, "was largely dependent on a compliant press that uncritically repeated almost every fraudulent administration claim about the threat posed to America by Saddam Hussein."[17]

Undoubtedly the Bush administration, like every other administration, tried to push public opinion toward support of its policies. Whether any or all of its claims about the Iraqi threat were "fraudulent," as MacArthur said, meaning intentionally deceptive, will be debated for a long time, probably without ever reaching definitive conclusions.

New Yorker correspondent Seymour Hersh, in an argument similar to MacArthur's, said, "If it is true that this administration deliberately, from the very beginning, understood that the best way to mobilize the American people was to present Saddam as a direct national-security threat to us, without having the evidence beforehand that he was, that's well, frankly, lying. That's the worst kind of deceit a President can practice." In such instances, argued Hersh, journalists should "hold public officials to the highest standard. . . . If we start saying that anything less than the highest standard is tolerable, we're really destroying democracy. Democracy exists on the basis of truth."[18]

Again, the issue may not be easily divisible into truth and lies. If the administration's information about Iraqi capabilities and intentions was debatable, should policy makers have waited until they had incontrovertible proof before acting, or might that have been too cautious? In the real world, as opposed to an idealized decision-making model, honest belief based on the best available evidence may have to suffice when it is time to act.

The news media should keep asking questions, but the primary impetus for debate must come from political leaders in Congress and elsewhere. Journalists are right to keep their skepticism intact and constantly demand that the government submit that best available evidence to public scrutiny. In the run-up to the Iraq war, the adequacy of proof was itself the subject of coverage (although arguably not enough), as Secretary of State Powell and others made the Bush administration's case to the world. Were the satellite photographs Powell displayed at the United Nations solid evidence of the existence of Iraqi chemical or biological weapons? The most sensible, if unsatisfying, answer was "Perhaps."

Journalists might not like that lack of certainty, but they have little choice other than to report what they can find out, while adding qualifiers that point out possible flaws in the official gospel and remind the public that news reports about such matters inevitably contain both fact and speculation. To contend that this approach means the news media are "compliant" and "uncritical," as MacArthur asserts, is unfair. If journalists keep pushing and digging, the truth may eventually emerge, although later than many would prefer.

Jessica Lynch

The saga of nineteen-year-old Army Pfc. Jessica Lynch was not just about the young woman from West Virginia; it was about symbols, public opinion, manipulation of information, and the uneven performance of the news media.

On March 23, Lynch was captured by Iraqi forces after her unit was ambushed in Nasiriyah, in central Iraq. The driver of the humvee in which Lynch was a passenger lost control when the vehicle was struck by a rocket-propelled grenade, and the humvee crashed into a truck. Two soldiers in the back seat were killed instantly, and the driver later died of her injuries in an Iraqi hospital. Lynch suffered extensive injuries, including multiple fractures. She was taken from the wreckage of the humvee to an Iraqi hospital, where she received medical attention that apparently saved her life. (American doctors later reported that at some time during her ordeal, Lynch may have been raped.) On April 1, U.S. special operations forces raided the hospital and took Lynch by helicopter to safety.

That is a good story as it stands, but it soon got better, at least according to *The Washington Post*. In a story headlined "She Was Fighting to the Death," *The Post* reported that during the Nasiriyah ambush Lynch "fought fiercely and shot several enemy soldiers . . . firing her weapon until she ran out of ammunition." Lynch, said the story, "continued firing at the Iraqis even after she sustained multiple gunshot wounds and watched several other soldiers in her unit die around her. . . . 'She was fighting to the death,' the official said. 'She did not want to be taken alive.' Lynch was also stabbed when Iraqi forces closed in on her position, the official said."

Deeper in *The Post*'s story was a disclaimer of sorts: "Several officials cautioned that the precise sequence of events is still being determined. . . . Reports thus far are based on battlefield intelligence, they said, which comes from monitored communications and from Iraqi sources in Nasiriyah whose reliability has yet to be assessed. Pentagon officials said they had heard 'rumors' of Lynch's heroics but had no confirmation."[19]

The uncertainty reflected in that passage might cause some news organizations to pause, but not *The Post*. Lynch-as-Rambo was an irresistible story, and after *The Post* ran it, it was picked up around the world.

Then came the problems. A few hours after the *Post* story appeared, the commander of the Army hospital in Germany where Lynch was being treated said there was no evidence of gunshot wounds. A *Post* story the next day quoted Lynch's father agreeing with that. A number of other stories in *The Post* about Lynch during the next ten days contained contradictory information from un-named military and medical personnel concerning the nature of her injuries. As *Post* ombudsman Michael Getler later wrote, the thin sourcing and the cautions about the details of what had happened to Lynch should have led to a toned-down account in the paper.[20] The fact that this was a great, uplifting story and that the basics of Lynch having been captured and rescued were true does not override the need to deliver better verified or at least better balanced reporting.

On June 17, *The Post* printed a two-and-a-half page story that provided details about the Lynch capture and rescue and that was implicitly a correction of the original story. Nevertheless, ombudsman Getler raised more questions about *The Post*'s handling of the story. Why had it taken more than two months to challenge the first version? What were the motives of those who leaked the story to the *Post* reporters? Was planting the story about Lynch's purported role in the firefight deliberate manipulation by the government?[21]

In July, the Army released part of a report about its investigation of the March 23 ambush, which made clear that Lynch had sustained all her injuries when the humvee crashed and had not been shot or stabbed. Her injuries were so severe that she had not been able to fight.[22]

That would seem to settle any debate about what had happened to Lynch, but questions remained about the U.S. government's role in packaging the rescue and the would-be legend. Lynch's rescue from a Nasiriyah hospital had been filmed with a night vision camera by the special operations team that extracted her. An edited version was delivered with considerable fanfare to the news media, which presented it to the public with even more fanfare. All this came at an opportune time for the Defense Department, counteracting gloomy news reports about U.S. forces being bogged down. BBC reporter John Kampfner determined that "her rescue will go down as one of the most stunning pieces of news management yet conceived. It provides a remarkable insight into the real influence of Hollywood producers on the Pentagon's media managers."[23]

The Defense Department had some defenders. *Time* magazine jumped into the debate, saying that Kampfner's BBC program "may be guilty of exaggeration itself, with its claim that the Pentagon manipulated information."[24] Victoria Clarke said the Defense Department had not pushed the rescue story. "We were waving them off, waving them down," she said, "and we actively discouraged any speculation about that story."[25] John Walcott, Washington bureau chief of

Knight Ridder Newspapers, supported Clarke's position, stating, "We have not been able to find anywhere in the record any military official overselling this as a heroic mission or claiming that it was done under heavy fire or anything of the sort."[26]

Journalist Mark Bowden, author of *Black Hawk Down* (the story of the calamitous 1993 battle in Somalia involving American troops), underscored the media's contribution to the Lynch saga, writing that a "heroic tale" was constructed, particularly on certain cable channels, "with slick packaging and great fanfare, with atmospherics borrowed from Hollywood. This is how the media works today, for better or worse. It happens without any prompting from the Pentagon; indeed, it would not have been possible for the Pentagon to stop the unspooling of *Saving Private Lynch*."[27]

Reasons that news organizations might overplay a story like this one can be found in the growing closeness between the news and entertainment industries, which sometimes live together under the same roof. A "great story" does not always mean great journalism, but rather "great" just in terms of audience appeal. When CBS News was trying to secure the first interview with Lynch after she returned home, a CBS executive dangled in front of her a package of enticements from Viacom, the conglomerate that owns CBS:

- A two-hour CBS News documentary
- A publicity campaign on various CBS programs
- An MTV special about Lynch's "dramatic coming of age"
- An MTV concert in Lynch's hometown, Palestine, West Virginia
- An hour on MTV2 hosted by Lynch and her friends
- A Country Music Television special, also produced in her hometown
- A two-hour made-for-TV movie (promised to be comparable in some ways to the series *JAG*)
- Discussions with Simon and Schuster about a book telling Lynch's story[28]

After the proposal was publicized, the network insisted that the news division maintains independence from the entertainment division. Critics said CBS was trying to buy a Lynch interview with a Hollywood-type deal.

Aside from such sideshows, the back-and-forth between journalists and government officials, as in the examinations of the Lynch case, is healthy because it keeps the various players from forgetting about the value of accuracy. Just as the news media's fondness for sensationalism should always be kept in mind, so too should the government's fidelity to the truth never be assumed. This caution is in order whether the topic at hand is hidden Iraqi weapons, Jessica Lynch's peril, or anything else.

The debate between the press and the Pentagon sometimes takes on a sharp edge. The job of media managers at the Department of Defense and elsewhere in the government is, after all, to manage the media. The job of journalists is to maintain a first line of defense through skepticism and reinforce that with aggressive, independent reporting. Government officials may occasionally be able to get away with exaggeration—and sometimes outright fabrication—but journalists will eventually catch up with them and the public usually will get a revised version of the story that is closer to the truth.

News consumers are generally aware that this is how the game is played. Most people get information from a number of sources and can cut through the fog of propaganda to figure out what is going on. Jim Hoagland of *The Washington Post* pointed out that "you find truth only in common sense—in the process of comparing and analyzing information yourself and then applying your life experiences to see where, how or even if it fits into the larger scheme of things." The public, said Hoagland, is quite capable of doing that.

Concerning the debate about the Bush administration's justification for the war, Hoagland wrote, "It is disingenuous to look back now and say that support for the war was built primarily on a belief that weapons of mass destruction would be found soon after battlefield victory." He cited a Time/CNN poll taken three weeks before the war that found that 83 percent of respondents said the most compelling reason to disarm Saddam Hussein was that "he has wantonly killed his own citizens." Addressing the Lynch story, Hoagland noted that only with "retrospective clarity" was it apparent that a less dramatic rescue effort would have sufficed. He pointed out that the rescue planners prudently suspected that Lynch might have been bait in a trap to attract rescuers, but he also noted that the Pentagon "failed seriously in the aftermath by not moving quickly and aggressively to correct a public record full of distortions and embellishments."[29]

Emerging from all this is recognition that precision in war coverage is, and will continue to be, elusive. It would be wonderful if the news media were able to see through every smokescreen and spot every gambit created by parties interested in influencing public opinion. But that is not going to happen, at least not until after the fact. So, as Hoagland noted, the public and journalists alike must rely on common sense and do the best they can to find a path through the information swamp.

Reporting and Saluting

Further complicating news media efforts to deliver accurate, objective information about the fighting was the tendency to succumb to boosterism. A mix of thoughtful patriotism and airheaded jingoism can affect news content—sometimes noisily, sometimes subtly. Consider, for example, pronouns. Should an American news organization refer to "U.S. troops" or "our troops"? "American forces" or "we"? This was particularly an issue for embedded reporters when discussing the units they were accompanying. Just as closeness made objectivity difficult, so too did it make semantics problematic.

Adding to this was what Terence Smith called "the cheerleading, can-do tone that infected too much of the reporting as U.S. forces advanced against an overpowered, overwhelmed enemy." The reporting, said Smith, seldom pointed out the miserable condition of much of Iraq's military, which had been pounded by U.S. bombing since the 1991 Gulf war.[30] The Pentagon, of course, was quite pleased to have Iraqi forces portrayed as formidable, because that made the American victory seem even more impressive. But the state of the Iraqi military—particularly whether it posed much of a threat to its neighbors or the United States—was an important topic because it was at the heart of whether the war was necessary. Important, but receiving skimpy coverage.

Some news organizations—Fox News, for example—operate on the principle that loudly proclaimed patriotism is good for ratings. In the short run, at least, that is probably true. Having Geraldo Rivera always ready to charge into battle will attract an audience of the curious, if not the knowledgeable. But after a while, when the public becomes more leery of government pronouncements about the state of affairs, news organizations that seem to be closely adhering to the government's line may find their audience going elsewhere, searching for more objective reporting.

A survey conducted in late June 2003 by the Pew Research Center found that Americans want a mix of objectivity and patriotism. Seventy percent of the survey's respondents said news organizations embracing a "pro-American" point of view was good, but when asked specifically if it is better for coverage of the war on terrorism to be neutral or pro-American, 64 percent favored neutral. Fifty-one percent said that news organizations generally "stand up for the U.S.," and 33 percent said news organizations are "too critical of the U.S." When asked about news media criticism of the military, 45 percent said that criticism helps keep the nation militarily prepared, while 43 percent said it weakened the country's defenses. This indicated growing concern about the role of the press and reflected a shift in opinion since the period following the Gulf war. When that same question was asked in March 1991, the public (by 59 to 28 percent) said press criticism of the military was a good thing.[31] Pew Research Center director Andrew Kohut said that Americans appear to not want propaganda, "but they want the media to be on our side, so to speak, giving you the sense that they have your values, your interests."[32]

The Iraq war offered more than the usual amount of boosterism, partly because the competition among news organizations—especially on television—was so fierce. Fox, in particular, staked claim to a niche audience of viewers with a fondness for red-white-and-blue coverage, and when it was successful imitators followed. News is not a unique product that only one network can offer, as a hit sitcom might be. Everyone is covering the same war, and so the different news products may be distinguished more by style than substance.

During the Iraq war, adding a hawkish tilt to coverage was endorsed by consultants who wield considerable influence over local television and radio news. A survey conducted by one prominent consulting firm, Frank N. Magid Associates, found little interest among TV news viewers in seeing coverage of protests against the war. The implied message from the Magid firm to its clients: covering the protests could hurt ratings. Magid's senior vice president Brian Greif said, "Obviously, you have to give both sides of the story, but how much time you devote to [protest coverage] and where you place it in your newscast becomes an issue."[33]

Coverage of anti-war activity was sometimes condescending, sometimes overtly critical. When Fox covered a large anti-war demonstration in New York in February 2003, it referred to "the usual protestors" and "serial protestors." CNN ran a headline about the event on its Web site stating, "Antiwar Rallies Delight Iraq." *New York Times* columnist Paul Krugman, writing a few days after the protest, argued that "for months both major U.S. cable networks have acted as if the decision to invade Iraq has already been made, and have in effect seen it as their job to prepare the American public for the coming war." He added that "U.S. media outlets—operating in an environment in which anyone who questions the administration's foreign policy is accused of being unpatriotic—have taken it as their assignment to sell the war, not to present a mix of information that might call the justification for war into question."[34] During late 2002 and early 2003, the administration's case almost always reached the public through the news media, while opponents—including members of Congress and retired four-star generals testifying before congressional committees—sometimes found themselves shut out from important news venues.[35]

The general tenor of some media organizations' approach to the war was reflected in advice from radio consulting firm McVay Media, which suggested that its clients use "patriotic music that makes you cry, salute, get cold chills! Go for the emotion. . . . Air the National Anthem at a specified time each day as long as the USA is at war."[36] McVay's Web site ran ads urging stations to

"Download free patriotic songs," and to use an animated American flag on their own Web pages.

Such patriotic gimmickry drowns out less popular voices and obstructs substantive coverage of the anti-war movement. This is important in shaping political debate, because if people are not exposed to arguments reflecting varied positions on an issue, they are less likely to think carefully about alternatives to the dominant viewpoint. The Internet provides a detour around mainstream media and can help disseminate diverse views, but it is a poor substitute for conventional news coverage as a vehicle for reaching a broad audience.

The growing concentration of media ownership in relatively few corporate hands will enhance the impact of programming that is slanted one way or another. For instance, radio stations belonging to Clear Channel adhered to a pro-war line and helped organize pro-war rallies around the country. Clear Channel owns more than 1,200 outlets in the United States.

Tilting the news attracted the attention of the head of the BBC, Greg Dyke, who warned against the "Americanization" of British media. He criticized the "unquestioning" approach of U.S. networks, saying that no news organization was "strong enough or brave enough" to stand up to the White House and the Pentagon. He said that the BBC was committed to "independence and impartiality," and if it were to adopt "the Fox News formula of gung-ho patriotism" it would risk losing the trust of its audiences.[37] Commenting on Dyke's critique, *The Economist* noted that Britain's news media were far from bias-free themselves, and cited the political leanings of Britain's newspapers, which are as competitive as America's television networks. Adopting varied positions on the war, the papers had been clearly partisan, presumably for the same reason that Fox had adopted its strategy: that's what news consumers wanted.[38]

Whatever the rationale behind partisanship in news and other media offerings in the United States and elsewhere, a steady diet of one-sided information will increase the intellectual isolation of those whose political outlooks are not "popular" at the moment. It is worth remembering that the media can stifle as well as foster debate.

Patriotism was an issue in NBC's mid-war firing of veteran correspondent Peter Arnett, who was covering the conflict from Baghdad. Arnett had given an interview to Iraq's state-run television in which he praised the Iraqi Ministry of Information and said, "Clearly, the American war planners misjudged the determination of the Iraqi forces. . . . The first war plan has failed because of Iraqi resistance, now they are trying to write another war plan."[39]

Arnett was wrong about the war plan failing, but arguably he was entitled to express his opinion. He was not alone in his criticism of the U.S. strategy; retired American generals were making some of the same points in interviews. But as journalist Carol Marin observed, "In the end, it all boils down to what appears patriotic and what does not. Or maybe it all boils down to money," meaning that NBC was afraid of losing a portion of its audience if it were to be perceived as the "unpatriotic" network.[40]

It should be noted, however, that the principal problem was Arnett's forum. Iraqi television was part of Saddam Hussein's government and Arnett must have known or should have known that his interview would be aired only if it was useful as propaganda. Such bad judgment on Arnett's part made him suspect in his employers' eyes as a news gatherer and in some of his audience's eyes as a source about what was happening in Baghdad.[41] That in itself was reason to fire him.

This case illustrates that there are some unwritten limits on dealing with the enemy in wartime. Staying in Baghdad, even while on a leash held by Iraqi government minders, was acceptable, and many who stayed—such as NPR's Anne Garrels and John Burns of *The New York Times*—delivered excellent reporting. But unnecessary cooperation with Saddam Hussein's regime, which is what Arnett's interview amounted to, exceeded those limits.

Keeping Secrets

Given the history of government deception in recent decades, using "national security" as a blanket justification for secrecy should always elicit a prompt and sharp response from journalists. When facing assertions of the supremacy of secrecy, wrote journalist Ted Gup, the press must "relentlessly challenge that supremacy and, as best it can, test the authenticity and credibility of those secrets that are revealed."

Even the shroud of secrecy wrapped around intelligence operations should be probed. The news media have a responsibility to explain how intelligence is used as a policy-making tool. The public should be reminded, said Gup, that "intelligence is not a hard science, but one that is riddled with nuance, that it requires interpretive skills, that it often produces contradictory or conflicting results, that it is not always immune to political pressures, and that, historically, its accuracy has been uneven. Context is everything; without it, stories about intelligence are misleading or outright unintelligible to many readers."[42]

The government sometimes provides the news media with "secrets"—usually those that support policy decisions. Journalists must decide how vigorously to test that information and how forcefully to seek more. The Bush administration's selective release of intelligence material about Iraqi weapons capabilities was an integral part of its strategy to win support for going to war. After the war, the information was criticized from within the intelligence community and elsewhere as being flimsy at best and purposeful deception at worst.

News stories about intelligence matters, particularly those about ongoing or recent operations, may be vulnerable to criticism for several reasons. First, the reporting may seem imprecise, with heavy reliance on anonymous sources whose information might be fuzzy and hard to verify. Second, the public may question the necessity of any news reports that could endanger operatives on the ground. In February 2003, *The Washington Post* ran a story headlined "Special Operations Units Already in Iraq," which reported that some U.S. forces had been deployed in Iraq and were looking for weapons sites, establishing a communications network, and probing for possible Iraqi defectors. *Post* ombudsman Michael Getler heard from readers who said they were "aghast," "appalled," and "incensed" that *The Post* would publish such a story. Getler observed that the readers' reactions were predictable and that the story should have addressed their concerns. He noted that the paper had checked with Pentagon officials before running it, and had dropped some geographic information at their request.[43] That background information was not, however, included in the original story.

A month later, *The Post* ran a story headlined, "U.S. Teams Seek To Kill Iraqi Elite," which said that CIA and military special operations teams were targeting Baath Party officials and Special Republican Guard commanders. This article might have raised concerns similar to those about the earlier piece, but it contained this sentence: "Provided with a detailed account of the contents of this article, U.S. government officials made no request to *The Post* to withhold any of the story's details from publication, as they have sometimes done in other cases involving ongoing covert operations."[44]

That sentence provides readers some insight into how the press and the government cooperate. Despite the public's belief that the news media sometimes behave irresponsibly when covering security issues, this story illustrates what often happens but is rarely mentioned—a news organization checking with the government to make certain that the operation and American personnel will not be endangered if the story appears. On the government's side, perhaps officials had no legitimate reason to ask that the story be altered or withheld. Or maybe they had their own reasons for wanting it published. They may have wanted to let Iraqi officials know that they were being targeted, which might have been incentive to surrender. Or perhaps there was no such operation; the material leaked to *The Post* may have been disinformation designed to make the Iraqi officials think covert units were tracking them when in fact no such units had been deployed. Whatever the truth may have been, this story is an example of the symbiosis that can exist between intelligence operations and news coverage.

On some occasions, news organizations refused requests that information be withheld. In late March, the Pentagon had asked that American networks not air video of U.S. prisoners of war being interviewed for Iraqi television until their families had been notified. CNN complied for several days, but then went its own way. CNN anchorwoman Judy Woodruff said on air that "now the Pentagon has asked that those interviews not be shown [at all]. But CNN has decided . . . that we would air brief audio of the POWs, because coverage of their treatment is an important part of the war in Iraq." Woodruff noted that CNN had first checked with the prisoners' families to confirm that they had been notified.[45]

This incident indicates that there is no clearly defined boundary between requests by the government that information be kept from the public for legitimate security reasons and attempts to manipulate public opinion by turning the flow of information on and off. News organizations must decide for themselves where that boundary is and should be wary about being too accommodating when the government argues for secrecy.

Sometimes news organizations withhold information for their own reasons rather than because of pressure from the government, and that too raises questions about the responsibility to disclose important news to the public. In April, as the heavy fighting was ending, Eason Jordan, chief news executive at CNN, wrote an op ed column for *The New York Times* telling of "awful things" that he had seen and heard during a dozen years of trips to Iraq. He said that CNN did not report these things because "doing so would have jeopardized the lives of Iraqis, particularly those on our Baghdad staff." Jordan cited the arrest and torture of one of CNN's Iraqi photographers and the threat made by Saddam Hussein's son Uday that he "intended to assassinate two of his brothers-in-law who had defected and also the man giving them asylum, King Hussein of Jordan." Jordan said that CNN did not report the former because of the danger in which it would have placed all the network's Iraqi employees, or the latter because of the danger to the translator the network had employed for the Uday Hussein interview. (Jordan did, however, privately warn King Hussein.) He related other horror stories of the Iraqi regime's use of torture and murder, all of which had gone unreported by CNN.[46]

The article stirred up responses reflecting two different views: that CNN had been cowardly by not reporting the truth about Saddam Hussein's regime, even at the risk of being thrown out of the country; alternatively, that CNN was courageous in putting compassion for their endangered Iraqi staff members ahead of their desire to break stories.

Mara Liasson of National Public Radio and Fox News said that CNN had made "a deal with the devil, trading for access." Writing in the *New York Post,* Eric Fetterman said that Jordan's approach "wreaks incalculable damage on all journalists' ability to be trusted by the American people."[47] John Burns of *The New York Times* said of Jordan's thinking that "the point is not whether we protect the people who work for us by not disclosing the terrible things they tell us. Of course we do. But the people who work for us are only one thousandth of one percent of the people of Iraq. So why not tell the story of the other people of Iraq? It doesn't preclude you from telling about terror."[48]

Media scholar and former journalist Alex Jones sympathized with Jordan, noting that "protecting your people always has to be the prime consideration." And Jordan himself, responding to the controversy, said, "I am at peace with myself knowing that I did the right thing and not put the lives of innocent people at risk."[49]

As David Folkenflik of *The Baltimore Sun* pointed out, Jordan's actions were not unprecedented. Folkenflik cited a 1959 essay by A. M. Rosenthal of *The New York Times,* who had been expelled from Poland because he wrote stories that angered the communist regime there. Rosenthal wrote that while he was safe at home, sources and others who had helped him remained behind at the mercy of the Polish government. "Every day in a communist country," wrote Rosenthal, "there are stories that a reporter must sit on for a while," partly because the Western reporter should "try to avoid getting people into trouble."

That seems to square with CNN's position. But Rosenthal had also written that "almost always it becomes possible to break a story eventually . . . after having made decently sure that the trail of sources has become weak and diffuse."[50] To do otherwise is to let the world's most vicious governments prevail. Journalists should hold such regimes to account by telling the world what they are doing. CNN and other news organizations will certainly find themselves again covering despotic regimes. When they do so, they might incorporate their concern for individuals' safety into a plan for reporting more of the truth.

Judging the Coverage

Donald Rumsfeld's observations about journalists' mood swings were echoed by some media critics. Howard Kurtz of *The Washington Post* wrote that newspapers "sometimes seem to be suffering from manic-depressive mood swings, with upbeat 'shock and awe' stories giving way to gloomy the-plan-is-flawed pieces, only to be supplanted by exciting battle-for-Baghdad reporting."[51]

Rumsfeld had implied that such shifts in the tone of coverage were due to journalists not knowing enough about what was really happening. Media scholar and former journalist Susan Tifft said, however, that Rumsfeld was adept at coming up with catchphrases about the news media that "label behavior in a way that makes it sound almost pernicious." She added that this "makes it very easy to discount what the press is doing."[52]

There was substance behind many of those reports that Rumsfeld didn't like. As Marvin Kalb noted, "journalists did not make up the anxiety felt by generals on the battlefield. They were told of their anxieties. . . . The media was doing its job by passing that information on to the American public."[53]

Rumsfeld and his critics raised valid issues. One reason for "mood swings" was the pace of the coverage—the constant emphasis on providing real-time reporting that had little room for dispassionate analysis or for alternative interpretations of what was happening. Mark Bowden suggested that a story could present a situation the following way: "Yes, at the moment it looks like supply lines are overextended because the advance has been so rapid. That could mean that troops could be strung out and could become vulnerable. It also could mean that we are on the verge of an amazingly rapid success."[54]

Another cause of mood swings was the Pentagon's own inflexibility in describing the state of the war. Given all the concerns about not repeating Vietnam-era mistakes related to the media and public opinion, it was surprising to see the Bush administration risk trapping itself with relentlessly upbeat appraisals that were sometimes based on wishful speculation rather than hard evidence. The Pentagon did little to knock down expectations about U.S. forces being greeted by thousands of surrendering troops and cheering civilians. When the best-case scenario turned out to be incorrect, the news media pounced, which should have surprised no one. Fortunately for Pentagon officials, none of them had claimed to see "a light at the end of the tunnel."

Rumsfeld was also critical of television news programs' propensity to use the same footage over and over, giving the impression that more was happening than was actually the case. Rumsfeld was especially irritated by stories about looting in Baghdad. "It's the same person walking out of some building with a vase," he argued. "You see it twenty times and you think, 'My goodness, were there that many vases?'"

The cable news channels are particularly susceptible to overusing material. The amount of time they must fill can exceed the amount of material they have available, so they replay images. The casual viewer, who tunes in briefly just to check the latest status reports, may not notice, but someone who watches for a longer time might think that the events being shown are continuous, not merely repeated video. David Zurawik of *The Baltimore Sun* noted that the news channels have a responsibility to recognize the power they have "to shape national perception" and therefore should "offer context to such red-hot images."[55]

Failure to define context is a persistent problem in war coverage. The shooting receives far more attention than the reasons behind the shooting. As media scholar Theodore Glasser observed, "Coverage trivializes war whenever strategy trumps substance, whenever 'who's winning' becomes more important than 'what's right?'"[56]

Not all American news consumers were oblivious to these matters. Some were uncomfortable with what they believed to be a pro-war slant to the coverage they were getting from U.S. media and so they looked elsewhere for their news. During the first three weeks of the war, the audience for BBC World News broadcasts carried by American PBS stations increased 28 percent. Based on e-mails it received, the BBC said that viewers were switching to the British newscasts for their "balanced and impartial" reporting.[57]

The BBC viewership was small compared to the overall broadcast and cable news audiences, but it was large enough to attract the notice of American news executives. The news audience has an expanding range of options to which it can turn, and as cable, satellite, and Web news offerings increase, the public will be more inclined to wander away from their traditional sources of news.

During the Vietnam war, Associated Press reporter Malcolm Browne, whose tough coverage from Saigon had won him few friends in the Pentagon, was asked by a U.S. Navy admiral, "Why don't you get on the team?" Even today, that question hovers over relations between the news media and the military.

The simple answer is that the job of the press is to report the game, not join the team. But patriotism complicates matters, and so the news media continue to search for terrain between the extremes of recklessness and cheerleading. During the Iraq war, most American journalists positioned themselves on that middle ground, but when the next war happens, they will have to find their way there again.

Meanwhile the new, uncharted territory of cyberspace is becoming increasingly important for those who cover and wage war.

CHAPTER FIVE

Cybernews, Cyberwar:

The Internet as Tool and Battleground

Here's the starting point for considering the influence of online news: the Internet has brought about the most significant changes in communications since the advent of television. In the long run, it will alter global society even more than television has because people use it not just to receive information but to send it and to connect with one another.

These characteristics were factors during the Iraq war. According to a study conducted by the Pew Internet and American Life Project, approximately 116 million adult Americans use the Internet, and 56 percent of them used the Web to get news and general information about the conflict and 55 percent used e-mail to communicate about the war. This does not by any means signal the abandonment of television. When Internet users were asked how they were getting most of their news about the war, television was far and away the first choice, but 17 percent said the Internet was a principal source of news. That might not seem like much, but it was up from 3 percent just after the 9/11 attacks.

Shortly before the war began, when there was less television coverage of Iraq-related matters, 26 percent said the Internet was a primary source of news about the possibility of going to war. This underscores the significance of the Web as an alternative to television when the major TV news providers—especially the big three broadcast networks—are delivering only their usual headline service. During the first five days of the fighting, the online news audience rose from roughly 29 million to 38 million Americans. Overall, 77 percent of the online Americans surveyed said that they used the Internet in some way to get information about the war.

Among the war-related uses of the Internet cited by respondents were look-ing for news (44 percent), checking reaction of the financial markets (23 per-cent), and getting background information about Iraq (15 percent). Smaller numbers took advantage of interactive aspects of the Internet: signing petitions for or against the war; reading or posting comments on a bulletin board or in a chat room; getting information about becoming involved politically. When In-ternet users were asked what was important about using the Net, they cited get-ting news from a variety of sources, getting up-to-the-minute information, getting points of view different from those of traditional news sources and the government, and exchanging e-mails about the war.

The survey results illustrate the growth of online news use and some of the reasons that this medium is increasingly popular. Convenient access—from cell phones and other devices, from offices and other locations—is important. So is the diversity of sources available. The Pew study found that while the Web sites of American television networks and newspapers attracted the largest audi-ences, substantial numbers of people visited U.S. government sites, foreign news organizations, alternative news sites, and sites maintained by pro- and anti-war groups.[1]

The Internet gives news consumers unprecedented independence as infor-mation gatherers. In pre-Web days, most information from governments, interest groups, and other sources was presented—and filtered—by the news media. Few citizens had the time, skill, or inclination to track down primary material, such as politicians' speeches, government reports, and the like. It was easier to let jour-nalists do all that work, put it in digestible form, and deliver it. But now, in the era of unmediated media, anyone can use the Web to easily access information from the White House, the Pentagon, other governments, NGOs, and even ter-rorist groups, as well as news media from throughout the world. During wartime, the Internet allows the individual news consumer to break free from parochial-ism and view events from many perspectives. Over time, that will have profound effect on the news business, governments, and the way the world works.

Some Online News Milestones

The Internet first made its wartime presence felt during the Kosovo war of 1999. On the first day of the NATO air strikes, CNN's Web site had 31 million page views. (The site had attracted 34 million page views in 1998 on the day the Starr Report about the Clinton-Lewinsky scandal was released.) During the first week of the conflict, CNN had more than 154 million page views. ABC-NEWS.com reported that its number of visitors increased more than 60 percent when the fighting in Kosovo began. Other news organizations' Web sites also saw large increases in the number of visitors. These were not just American news consumers; Yugoslavs' visits to CNN.com rose 963 percent.[2]

Non-news organizations also used the Web to deliver their messages. On its site, the Ministry of Foreign Affairs of the Federal Republic of Yugoslavia offered articles from the Yugoslav press and favorable foreign news stories. It also featured a section labeled "NATO Aggression," presenting the Serb view of the enemy. On the other side of the war, the NATO Web site showcased its version of events, supplemented with downloadable video from bombing runs. The United Nations High Commissioner for Refugees provided Web information about the status of Kosovar refugees and the International Criminal Tribunal for the Former Yugoslavia included records of its war crimes trials. Smaller special interest organizations and even individuals developed Web sites as forums for their own information and viewpoints.[3] Families posted information about missing relatives, relief agencies solicited donations, and people trapped in the midst of the fighting e-mailed descriptions of what was happening.

During the fighting in Kosovo, as had been the case during the Clinton impeachment, the Web proved itself particularly valuable as a repository of background material that the public normally has trouble finding. Regular news formats had no room for this kind of content and other traditional information sources such as research organizations and libraries had no way to get it to the mass public. Web sites, with their virtually infinite capacity, have plenty of space for maps, timelines, biographical sketches of major players, documents, and other such resources, as well as interactive and multimedia materials. For anyone who wants to go on line to get it, a mammoth reservoir of information awaits.

A major event in the coming of age of the Internet as a news medium was the attack on the United States on September 11, 2001. Television, of course, was dominant, with people reflexively turning on the TV as word spread about what had happened. This was partly because of the role of the network anchormen, who in times of crisis are de facto national leaders, offering reassurance through their calm electronic presence. But people also turned to the Web in unprecedented numbers. On 9/11, 30 million people went online solely for news coverage. During each day of the week following the attack, there were almost 12 million unique visitors to news Web sites.[4] On the 11th, washingtonpost.com saw three times its previous record number of users (which had been set on the day after the 2000 election).[5]

This expansion of audience was accompanied by problems, mostly relating to the still-evolving online technology. Especially during the first hours after the attacks, major news sites were overwhelmed by the number of visitors and were often inaccessible. In the days after the attacks, the flood of visits to news sites became more manageable, and people turned to the Web as an alternative and a supplement to other news sources, mainly television. Rather than accepting the measured pace and often repetitive content of the networks' coverage, Web users could search for what they wanted to know. They also could take advantage of the graphics and background information that the Web offered, and they could scan international news sites to see how the rest of the world was interpreting events.

Aside from the surges in Web use sparked by events such as the Kosovo war and 9/11, the Internet is being increasingly integrated into the public's information-gathering patterns. One attribute of Internet news that was apparent during the Iraq war is the encouragement it implicitly gives its users to search more widely for information. American Internet users did just that. During March 2003, American visits to Al-Jazeera's Web site increased to 1,037,000 from 79,000 in February. BBC World Service's site received more than five million American visitors in March, a 158 percent increase over the previous month. Reuters, which had added streaming video to its site, saw a 72 percent increase during the month, reaching more than two million visits. While the popularity of international news sites grew, the two leading American news sites, CNN and MSNBC, remained the most popular, with 26 million and 24 million visits respectively, both of which were up about 24 percent from February.[6]

This exploration of sources by news consumers was stimulated by numerous reasons, such as unhappiness about perceived bias in American news coverage and the ease of access the Web provides. Although the numbers of visits to news sites dropped after the war, patterns of getting the news are changing. Some people may go back to sites they encountered and liked during the war. When the next international crisis arises, those people who sampled news from elsewhere in the world in 2003 may do so again. Ideally, American news organizations will take note of this new level of competition and respond by improving their own coverage. Realistically, however, that probably will not happen until the international providers make more substantial inroads into the American audience.

This is one of those quiet but substantial changes illustrating that globalization is much more than simply a theory.

The Online News Product

The beginning of the U.S. invasion of Iraq was accompanied by a surge in Internet use. Akamai Technologies, Inc., which manages traffic for more than a thousand of the Web's largest sites, reported that on March 20 it served almost 25 billion requests for Web pages, roughly double its average day the month before. That same day, traffic to Yahoo's news section increased 600 percent. CNN.com had nine million visitors, 200 percent more than its average traffic over the previous month. Other television news organizations' sites also showed big increases in traffic, due in part to on-air promotion of the Web offerings. [7]

Internet news is constantly tested by comparing it to other media. One of the proponents of Web journalism, Mitch Gelman, executive producer of CNN.com, said, "You're combining the speed of television with the depth of print."[8] Kinsey Wilson, vice president and editor-in-chief of USAToday.com, pointed out that most of the audience turns to television first to get the narrative story line, but eventually "television starts to loop back on itself and repeats the narrative over and over again." That is when the Web gains the advantage, he noted, because "the best sites can move quickly to develop a story in multiple directions, add depth and detail, and give readers their own pathways to explore."[9]

Barb Palser of Internet Broadcasting Systems, Inc. wrote that administrators of Web news sites had plenty of lead time (as opposed to the suddenness of 9/11) to design their pages, establish ties with content providers, and make certain their servers could handle an expanded audience. Although the Internet did not steal a large audience from other media, said Palser, the Web advanced significantly in three respects: journalists became more accustomed to using the Web as a tool in doing their work; news sites designed innovative formats; and the technology worked, carrying the burden of unprecedented levels of use.[10]

The most important breakthrough in terms of technology-dependent content was the increased use of streaming video and audio. These images and sounds are not separate large files that must be downloaded but rather feed continuously for watching and listening as they arrive, turning the computer monitor into something like a television set. For decent "reception," a broadband connection is necessary.

Television-level video quality is what the news audience is presumed to want if it is to embrace a new medium. The demand for video clearly exists. Real-Networks, which provides the streaming technology used by many Web sites, reported that on September 12, 2001, 11.4 million streams were accessed on its site, compared with the previous daily average of 1.5 million streams.[11] During the first three days of the Iraq war, MSNBC provided an average of 8.7 million video streams per day, up from a pre-war average of fewer than 500,000 per day, and by war's end had provided more than 60 million video streams.[12] MSNBC's streaming product, like those of most news organizations, was free, but some providers—including Yahoo, ABC, and CNN—provided an additional, premium service, charging about five to ten dollars a month. ABC News Live, which began providing for-fee content a week before the war began, may evolve into a full-scale Internet news network, a Web version of CNN.[13] With so much free material available, the services that charge must provide innovative content and take full advantage of the at-work audience—people in offices and elsewhere without access to television but with computers where they can check the status of big stories. For them, a "television-quality" product on line may be worth a few dollars a month, especially during a major event such as a war.

The proliferation of broadband connections will make the Internet even more competitive with other media because it provides vastly improved video quality, as opposed to that received through dial-up connections. By April 2003, 35 percent of the U.S. Internet population was using this technology. This is a substantial number, but far behind the 82 percent of Hong Kong's Internet users who rely on broadband, and trailing even the Internet populations of France, Spain, and the Netherlands. Other countries, such as Great Britain have been rapidly catching up, with the number of its broadband users more than tripling during a year's time.

Nielsen NetRatings research found that the broadband boom was leading to Internet users "spending significantly more time on line, using the Web more often, and visiting more Web sites than their slower dial-up counterparts." If the current trend continues, the majority of Internet users will soon be using broadband, which means that most Web material will be designed for them.[14] High quality online video and audio will be standard, and combined with the other attributes of the online world, this could be a big step toward making the Internet a more dominant medium. The next war will be watched on line, with real-time and supplemental news flowing through cyberspace to the public.

During the Iraq war, many online news offerings were available from free as well as from for-fee sources. These providers included the following:[15]

- Reuters Raw Video, which featured many vivid scenes from the battlefield with minimal editing and narration
- WashingtonPost.com video, provided by Travis Fox, who used a small digital video camera and showed how a newspaper can independently compete in a video-oriented medium
- Audio reports, particularly from newspaper correspondents
- Photo essays from Knight-Ridder, the *St. Petersburg Times,* the Canadian Broadcasting Company, and others
- CNN.com's urban warfare animation, which illustrated the perils of street-by-street fighting

- The BBC's Military Facts File, which provided details about American, British, and Iraqi military capabilities
- Interactive maps and other information, some of it narrated, provided by various news organizations

There was much more, changing throughout the war as journalists and Web designers experimented with the capabilities of the evolving medium. Aside from all the bells and whistles, the end product was a huge reservoir of background information and timely reporting—as much as any news consumer could want. In terms of the quantity of coverage, this was an unprecedented performance. As for quality, selective news consumers could find much intriguing and useful material.

Bloggers, Smart-Mobbers, and Others

While major news organizations displayed their electronically enhanced muscle, the Internet was also being put to use by individuals and groups that might not have been heard otherwise. The Internet is a great amplifier; anyone can reach millions instantly through a Web site and e-mail. The cost is negligible. For equipment a laptop computer will do nicely, and the principal challenge is simply being found amidst the vast Internet universe.

The Iraq war saw the rise of the Weblog or blog—first-person accounts of events, often laced with a dose of opinion. Weblogger "L. T. Smash," a U.S. military officer in the middle of the fighting, filed reports such as this: "Saddam fired a couple of those Scuds that he doesn't have at me this afternoon. He missed." During the first week of the war, his Web site was getting 6,000 visitors a day. Howard Kurtz of *The Washington Post* characterized the blogs as "idiosyncratic, passionate, and often profane, with the sort of intimacy and attitude that are all but impossible in newspapers and on television."[16]

The Webloggers' ranks included Baghdad residents, politicians, and journalists who supplemented their regular output with more personal reporting. Of the latter, Kevin Sites of CNN attracted a considerable following to his personal blog, but he was soon told by his bosses to devote his energies to reporting just for them. A member of the British parliament kept an on-line diary that reflected the difficulties of shaping war policy.[17] Freelance online journalist Christopher Allbritton requested donations to pay for him to travel to Iraq, and he quickly raised $14,000. When he got to Iraq, he put his donors on a premium e-mail list, so they received his stories early and also got extra reports and photos.[18]

Central Weblog sites, such as warblogging.com, pulled together material from individuals and items from international news media. Some media organizations set up sites specifically for correspondents' observations, among them *BBC Reporters' Log, CNN.com On the Scene,* and *MTV: Gideon's Journeys in Kuwait.* Sites were also dedicated to Weblogs from the families of military personnel, and the Poynter Institute offered sites with collections of coverage from local U.S. news organizations and discussions of journalism issues.[19]

The blogging sites shared a problem with much of the rest of online news: There is so much information on the Web that only the most dedicated news-seeker will reach most of the sites. Despite the overall rise in using the Internet to get news and the publicity given to the warblogs, only 4 percent of American Internet users went to the blog sites. The Pew Internet and American Life Project found that these sites were most popular among Internet users under the age of thirty.[20] Although the audience was relatively small, Nielsen estimated that 500,000 blogs were active in early 2003.[21]

During the war, several issues arose about webloggers' content. News organizations weren't certain about whether to restrict or encourage expressions of opinion from their reporters on their personal blog sites. The concern was that the audience for the reporter's regular and presumably objective coverage would read that same reporter's weblog and assume that the opinions expressed there affected his or her mainstream coverage. Even with a disclaimer that the weblog was a personal statement, not cleared by the parent company, some editors and producers were uncomfortable with their correspondents reporting in one venue and editorializing in another. Other news executives saw the blog sites as a useful, supplemental news product and free advertising, and so encouraged their bloggers. The question of editorial independence versus editorial control will probably receive more attention as the audience for blogs grows.

Another issue is the potential for manipulation by creating phony bloggers whose real mission is to deliver propaganda. During the war, questions arose about the identity of "Salam Pax," whose blog detailed daily life in Baghdad. Were his endorsements of the need for change in Iraq genuine, or was this a CIA creative writing exercise? After the war, *Guardian* reporter Rory McCarthy tracked down the blogger—a twenty-nine-year-old architect and computer geek living with his parents in Baghdad.[22] The online writings of Salam Pax were collected in his book, *The Clandestine Diary of an Ordinary Iraqi,* and offer intriguing insights into life in Baghdad before, during, and immediately after the American invasion.

The "Salam Pax" case aside, the Web does lend itself to deception, partly because anyone can devise an identity that is more convenient than true and make it part of virtual reality (at least for the short term). Although there is no foolproof defense against being manipulated in this manner, keeping skepticism intact is useful when exploring the Web.

In addition to surfing through the blogs, news consumers looking for coverage provided by sources other than established media organizations could find plenty of war-related reporting on the Web. AlterNet.org, for example, was founded in 1998 and maintained a database of more than 7,000 articles from more than 200 sources, including many with an anti-war perspective. Altmuslim.com was established to offer "an open-source compilation of issues affecting the Muslim world" and to help prevent the American Muslim community from being marginalized. In the past, the reach of such efforts would be severely limited by the costs involved in publishing and distributing an on-paper product. Now the task is not to raise funds but to get the word out that the Web site is open for visitors.

Through these and other independent sources, the Internet dilutes the potency of governments' "official news," which is most effective when it faces minimal competition. The voices of the Web are obstacles to the flow of propaganda because they provide so many information options. As Owen Gibson of the *Guardian* put it, "when you have a worldwide depository of millions of points of view, the propaganda war becomes a lot harder to win."[23]

Web-based information sometimes provides an alternative to conventional wisdom, and it may enhance citizens' ability to do more—to participate in political action. In the world of computers, "networking" has several meanings, and one of the most literal is the formation of virtual communities that may be unified by one or several shared characteristics, including political outlook.

Online activist Eli Pariser has been a pioneer in using the Web as an anti-war networking tool. After the September 11 attacks, Pariser supported a limited police action in Afghanistan rather than a full-scale war in response. By October 9, 2001, he was able to deliver a petition calling for that response that was 3,000 pages long and had a half-million signatures, all gathered over the Internet.[24] The petition did not bring about changes in U.S. policy, but the effort to get those names gave rise to a new kind of anti-war activism.

Pariser's project, 9–11 Peace.org, joined forces with MoveOn.org, which had been formed in 1998 by people opposed to the impeachment of Bill Clinton. MoveOn is part of an international network of more than two million online activists, and has worked to rally support for causes such as campaign finance reform and environmental protection. It has a political action committee that disbursed more than $3.5 million during the 2002 election campaigns.

MoveOn is one of the best examples of "smart mobbing"—galvanizing mass political action electronically. It organizes "virtual" demonstrations; instead of marching on Washington, advocates of a particular policy position deliver a carefully planned deluge of phone calls, faxes, and e-mails to legislators. When physical presence is required, a real march on Washington or other action can be coordinated on line.

The Web was loaded with sites representing opposition to the war. PoetsAgainsttheWar.org posted 13,000 anti-war poems on its site. WaketheWorld.org made hundreds of free anti-war posters available for downloading from its site and was attracting upwards of 80,000 visitors each week.

Sites in support of the war, although outnumbered, could also be found. Among them were Americans for Victory Over Terrorism (avot.org, led by William Bennett) and Patriots for the Defense of America (defenseofamerica.org). Although perhaps lonely amidst the passionate proselytizing about the war, organizers of sites such as opendemocracy.net encouraged balanced debate about war-related issues. [25]

Meanwhile back in Baghdad, before the war there were about 30 Internet cafes in the city, almost all state-owned and monitored. When trying to reach sites not favored by Saddam Hussein's government, would-be surfers would see "Access Denied" pop up on their screen.

At war's end, the Internet was quickly put to work as part of Iraqis' reconstruction efforts. A few weeks after U.S. forces had seized the city, some Iraqi entrepreneurs invested $10,000 in a satellite modem and a year's worth of e-mail service, set up a diesel generator to provide a relatively stable supply of power, and acquired a pistol to protect their venture. At their Internet café, they charged five dollars per hour for Web browsing, $1.25 for outgoing e-mails, and 75 cents for incoming ones. When they closed up each night, they hid the modem and took their daily gross of about $60 with them.[26] Soon, providing ways to reach out to the world through the Internet, satellite phones, and other high-tech means became a thriving industry in Iraq.

This, on a small scale, was freedom.

Warfare on the Internet

In addition to transforming conventional weaponry, advanced technologies are expanding the structure of war. On one level is information warfare, which is defined by defense analyst Bruce Berkowitz as "whatever you need to get to the end of your decision cycle before your opponent gets to the end of his." This involves collecting, analyzing, and using your own information and preventing your opponent from doing the same.[27] Some of this involves traditional intelligence gathering and operational planning; the newest technologies enhance and speed up these processes.

At a more aggressive level are cyberwar and netwar. Cyberwar is defined by defense and information experts John Arquilla and David Ronfeldt as "a comprehensive information-oriented approach to battle that may be to the information age what *blitzkrieg* was to the industrial age." They define its sibling, netwar, as "a comprehensive information-oriented approach to social conflict." Cyberwar, say Arquilla and Ronfeldt, will usually involve formal military forces, while netwar will be fought primarily by nonstate, nontraditional forces.[28] Internet expert Robert Lemos adds that "cyberattacks come in two forms: one against data, the other on control systems. The first type attempts to steal or corrupt data and deny services. . . . Control system attacks attempt to disable or take power over operations used to maintain physical infrastructure, such as 'distributed control systems' that regulate water supplies, electrical transmission networks and railroads."[29]

Both cyberwar and netwar involve disrupting or destroying information and communications systems. Disrupting communication is not a new tactic. Messengers have been intercepted and signal posts attacked since the earliest wars. As society generally, and war making in particular, have become more dependent on communication and information technologies, disruption—overt and covert—has become more important and more sophisticated. The work of code makers and code breakers, for instance, is an important element of the history of warfare. Among the examples of code breaking that had significant effect on past conflicts was the Zimmermann telegram, which contained Germany's secret 1917 proposal to Mexico to join in war against the United States and in return be able to reclaim Texas, Arizona, and New Mexico. British Intelligence had cracked the German code and arranged for the message to be presented to President Woodrow Wilson. This case of covert information warfare helped nudge the United States closer to entering the First World War.[30]

In recent conflicts, cyber/net tactics have ranged from harassment to more damaging efforts. During the conflict in Kosovo in 1999, hackers targeted NATO's computers, but damage was primarily limited to making Web pages more difficult to access.[31] In 2000, after Hezbollah seized three Israeli soldiers in Lebanon, pro-Israeli hackers created a Web site to host efforts to electronically attack Hezbollah. Terrorism scholar Maura Conway found that "within days, Hezbollah's site was flooded by millions of 'pings'—the cyber-equivalent of knocks on the door—and crashed." Sites run by Hamas and Palestinian groups were also flooded. Even when Hezbollah increased its server capacity, the pro-Israeli attacks were successful. In a counterattack, or "cyberjihad," pro-Palestinian hackers hit more than 200 Israeli-related sites.[32]

During the run-up to the 2003 war in Iraq, American cyber warriors launched a wave of e-mails and cell phone calls to Iraqi officials urging them to break with Saddam Hussein. Meanwhile, U.S. aircraft dropped tens of millions of leaflets urging Iraqi soldiers and the general population to turn against Saddam's regime, and an airborne broadcasting center, the "Commando Solo" aircraft, beamed its message into Iraq.[33]

The Iraqi leadership was not defenseless against this kind of attack. It had a system in place to censor e-mail coming into the country, and whenever a heavy volume of U.S. messages began arriving, Internet access throughout Iraq would suddenly suffer a "service outage."[34] Iraqis also began visiting U.S.-based Web sites that contained information about psychological warfare and offered links to military sites. Although it was impossible to identify precisely who from Iraq was accessing the American sites (the visits were emanating from Internet addresses assigned to Iraqi Internet services), it was assumed that at least some of the visitors were working for Iraqi intelligence. This illustrates two sides of the Internet's role: It is easy for anyone, including enemies, to acquire information on line, but it is also easy to manipulate Web visitors through disinformation or other means. James Lewis of the Center for Strategic and International Studies suggested that before the war began U.S. government sources may have planted stories about electromagnetic pulse bombs and other frightening high-tech weapons, knowing that the stories would be read on the Web and hoping that they would intimidate Saddam Hussein's military.[35]

Conflict on the Internet involves weaponry unlike that found on the conventional battlefield. Viruses, worms, and direct attacks can wreak electronic havoc that in an increasingly computer-dependent society must be taken seriously. Potent viruses and worms first attracted attention during the 1980s, and in 1998 intruders infiltrated and took control of more than 500 military, government, and private computer systems in the United States. First thought to be Iraqis, the culprits turned out to be two California teenagers. This incident gave the Defense Department and others reason to ponder what more sophisticated adversaries might accomplish. In July 2001, the Code Red worm infected Microsoft operating systems and targeted the White House Web site. It was designed to assault the White House with so many e-mail messages that it would knock the site off line. The White House defended itself by changing its numerical Web address. Additional worms and viruses have continued to appear, with some neutralized quickly while others have caused substantial financial damage as they paralyze or destroy computer services.[36]

Code Red had a $2.6 billion impact, and the biggest bug (as of mid-2003), the 2000 Love Bug, caused $8.75 billion in damage. The worldwide costs of cyber attacks rose from $3.3 billion in 1997 to an estimated $12 billion during the first eight months of 2003.[37]

Quick moves and countermoves are common when cyberwarfare is underway. Whoever has the best technology at the moment prevails, but that advantage may be short lived. The other side might come up with something new and leap into the technological lead.

A subsidiary enterprise of cyberwar is cyberterrorism. Information warfare scholar Dorothy Denning defined this as "a computer-based attack or threat of attack intended to intimidate or coerce governments or societies in pursuit of goals that are political, religious, or ideological. The attack should be sufficiently destructive or disruptive to generate fear comparable to that from physical acts of terrorism. . . . Depending on their impact, attacks against critical infrastructures such as electric power or emergency services could be acts of cyberterrorism. Attacks that disrupt nonessential services or that are mainly a costly nuisance would not." Although many cyber attacks, said Denning, "have caused billions of dollars in damage and affected the lives of millions, few if any can be characterized as acts of terrorism: fraud, theft, sabotage, vandalism, and extortion—yes, but terrorism—no."[38]

Denning also has noted that terrorists may stay with conventional tactics once they discover that cyber attacks are harder to control to achieve a desired level of damage. Also, these measures may not have the emotional impact of more conventional attacks, such as those using explosives.[39]

Cyber security expert D. K. Matai said that such tactics move up to the level of terrorism if they are "command and control" attacks on targets such as water, power, telecommunications, and transportation facilities, which could lead to fatalities. Computer viruses have been used in Spain and Japan to trick mobile phones into dialing local emergency numbers, which could flood and disable a 911 or similar system.[40] In 1998, Tamil guerrillas flooded Sri Lankan embassies with thousands of e-mail messages, which some U.S. intelligence officials said was the first known terrorist attack on a government's computer system.[41]

Terrorist organizations also have quietly infiltrated cyber infrastructure. Dorothy Denning reported that in Japan, a police department's software system used in tracking police vehicles turned out to have been developed by members of the Aum Shinryko cult, which released poison gas in the Tokyo subway in 1995, killing 12 people and injuring 6,000 others. Cult members had also developed software for ten other government agencies and at least 80 Japanese businesses, perhaps including "sleeper" flaws that could facilitate future cyberattacks.[42]

Whether conventional hacking or more surreptitious tactics are being used, cyberterror attacks are likely to increase in number, sophistication, and potency. In October 2002, the 13 computer servers that manage the world's Internet traffic sustained "distributed denial of service" attacks that apparently were designed to swamp the servers with so much data that they would collapse. Safeguards built into the system worked well enough so that ordinary Internet users experienced no slowdowns or outages.[43] A few hours after the first hour-long attack, another began, targeting specific global servers. It, too, was unsuccessful in terms of disrupting the world's Internet operations.[44] But these attacks left behind some important unanswered questions: Were these cyber attacks launched by hackers just for the sake of hacking, or were they the work of terrorists or even a government's operatives? Were these episodes isolated or were they preliminary probes before attempting a more overwhelming attack?

Despite the growing list of occurrences of various kinds of cyberconflict, debate continues about how serious a problem this is. Some people see it as a substantial threat to national security, while others dismiss the matter as being merely the stuff of science fiction.

In June 2002, *The Washington Post* reported that unknown Internet visitors had been probing digital systems of utilities and government offices in the San Francisco Bay area, including remote control of fire dispatch services and pipeline equipment. Suspicion centered on Al-Qaeda, and a CIA memorandum cited the group's interest in cyberterrorism, presumably to accompany and amplify conventional attacks.[45] Ron Dick, former director of the FBI's National Infrastructure Protection Center said that he was concerned about "a physical attack on a U.S. infrastructure, which is combined with a cyber attack that disrupts the ability of first responders to access 911 systems, that disrupts our power grids such that, again, first responders can't respond to an incident."[46] A February 2002 letter to President Bush signed by 54 scientists and defense and intelligence experts warned that America was "at grave risk of a cyber attack that could devastate the national psyche and economy more broadly than did the September 11th attack. . . . All that is required is a small group of computer scientists, a few inexpensive PCs, and Internet access. Even the smallest nation-states and terrorist organizations can easily muster such capabilities, let alone better-organized groups such as Al-Qaeda. Many nations, including Iran and China, for example, have already developed cyber-offense capabilities that threaten our economy and the economies of our allies."[47]

John Arquilla said that although the threat might not be imminent, he saw parallels to air power and to the early theories about attacks from the air being used without first engaging armies or fleets. Cyberwar was like that, said Arquilla, and "you don't even need a military in order to engage in this fashion. So it is a form of strategic bombardment."[48]

On the other side of the question are those such as *Washington Monthly*'s Joshua Green, who argued that there is "no instance of anyone ever having been killed by a terrorist (or anyone else) using a computer. Nor is there compelling evidence that Al-Qaeda or any other terrorist organization has resorted to computers for any sort of serious destructive activity."[49] Former deputy secretary of defense John Hamre contended that "terrorists are after the shock effect of their actions and it's very hard to see the shock effect when you can't get your ATM machine to give you twenty dollars." Scott Charney, Microsoft Corporation's chief security strategist agreed with Hamre, saying that "an attack on the Internet will not yield the kind of graphic pictures that you saw on 9/11." Charney also pointed out that "it's not as easy to take down the Internet as some might believe. There's a lot of redundancy, a lot of resiliency in the system."[50] A survey of possible cyberterror targets—electricity, surface transportation, water, energy, financial, and information technology—found that industry officials in each area took the threat of cyber attacks seriously but were more concerned about physical attacks, partly because they were not certain what cyber attacks would target and how they might work.[51] Despite the uncertainty about the potency of cyber attacks, an August 2003 Pew survey found that 49 percent of Americans "fear that terrorists might cripple American utilities . . . or its banks and major corporations through cyber attacks."[52]

On another level, terrorist groups operate on the Internet through their Web sites. These sites are sometimes well established and easy to find and sometimes, like their proprietors, are constantly on the move and hard to locate. Al-Qaeda has used numerous sites, including alneda.com, which has been knocked off the Internet numerous times by service providers when they discover whom it belongs to. Alneda.com was able to stay on line as an Internet parasite—embedded within another Web site without that site owner's knowledge. Around the time of the Iraq war, it had lodged itself within the subdirectory files of a legitimate Dutch business Web site. It was discovered and so it moved on. Al-Quaeda has reportedly used its site to urge terrorists to kill Americans and others and has posted messages purportedly from Osama bin Laden.[53] Even when Al-Qaeda is on the run, the Internet gives it a vehicle for getting instructions to its followers and propaganda to a larger public.

When U.S. forces moved into Baghdad, Al-Qaeda's reaction came via its Web site and said, in part: "Yes, we cannot deny that we were surprised at the ease of their entrance into Baghdad, and that there was no resistance whatsoever by the city. So far we are incapable of understanding where the tens of thousands of the regular troops disappeared to overnight. We expected some resistance. . . ." The message urged Iraqis to turn to guerrilla warfare as "the most powerful weapon Muslims have, and it is the best method to continue the conflict with the Crusader enemy."[54]

Anti-terrorist organizations are also using the Web. Internet Haganah relies on the Internet expertise of Israelis, Americans, and others in "confronting Islamist terrorism and its supporters on line." Describing itself as "an online counter-insurgency," Internet Haganah spotlights sites that endorse terrorism and tries to have them shut down by alerting law enforcement agencies and Internet hosting services.[55]

Against this backdrop of terrorists' use of the Internet and the ongoing debate about how seriously the threat of cyberterrorism should be taken, the prospect of cyber attacks has spurred development of a substantial industry to help guard against cyber attacks. Private companies specializing in Internet security are in demand, and the U.S. government, as part of its homeland security efforts, issued a report about the steps it needs to take. The report, "The National Strategy to Secure Cyberspace," warns that "our economy and national security are fully dependent upon information technology and the information superstructure." The report calls for a federal plan to address protection of critical infrastructure, devise crisis management measures, and upgrade federally supported research to assist private sector protective measures.[56]

The report also defines the national security issue concerning cyber security: "America's cyberspace links the United States to the rest of the world. A network of networks spans the planet, allowing malicious actors on one continent to act on systems thousands of miles away. Cyber attacks cross borders at light speed, and discerning the source of malicious activity is difficult." The report acknowledges that "the speed and anonymity of cyber attacks makes distinguishing among the actions of terrorists, criminals, and nation states difficult, a task which often occurs only after the fact, if at all."[57]

American responses to the threat of cyber attacks include military efforts, such as those of the U.S. Strategic Command's Joint Information Operations Center, which "is responsible for the integration of Information Operations into military plans and operations across the spectrum of conflict." This involves "actions taken to affect adversary information and information systems while defending one's own information and information systems."[58]

Measures undertaken by civilian and military bureaucracies are the early stages of what will presumably be a permanent function of government—protecting public and private interests in cyberspace. Even though many questions exist about the extent of the cyberwarfare threat, it makes sense to treat this as a serious security issue. For journalists, that may require adjustments in the ways in which communication is viewed; it must be looked upon as a tool of war as well as an information tool.

The Internet's traffic is estimated to double every hundred days. This underscores its expanding importance and the need for it to be reevaluated as a topic for news coverage. This medium that journalists, among others, rely on so heavily can be used as a weapon, and so should be a subject of news coverage in that context as well as others.

The Internet's future as a battleground may be uncertain, but no doubt exists about its continuing expansion as an information medium. Among its most significant functions is to serve as one of the tools being used throughout the world to give voice to those who long have not been heard.

CHAPTER SIX

The Din of Many Voices:

The Impact of Globalized News

Although the United States proved in Iraq that it retains absolute dominance in military power, this war marked an end to the near-monopoly in global news that American and other Western media had long enjoyed. New voices emerged, competing for audiences throughout the world by offering news shaped by varied interests and perspectives.

Competitive war coverage was just the latest step away from U.S. cultural hegemony. Earlier evidence of this gradual change could be seen on European television, where American entertainment programs had long been among the most popular shows. During the late 1990s, entertainment programming on European stations was increasingly homegrown or drawn from sources other than American studios. Between 1996 and 2002, the number of hours of American programming dropped by 26 percent in Spain, 17 percent in Germany, and 9 percent in Italy, according to the research organization Essential Television Statistics. In Britain, by 2003 the top 50 television programs were British-made.[1]

As for news programming, although Qatar-based Al-Jazeera has attracted the most attention, it is just part of a movement toward diversity. An example of this broader trend is the effort within France to establish an international all-news channel to compete with CNN and the BBC. An official of the French broadcasters' union endorsed this, saying, "Television and news are controlled by the United States, which is engaged in an extraordinary disinformation campaign." President Jacques Chirac supported the plan, which would involve some government management and funding and would target French-speaking populations around the world. Particular emphasis would be placed on reaching the Muslim world, where France hoped to parlay its opposition to the Iraq war into greater long-term influence.[2]

Chirac's stance underscored the importance of political as well as economic factors in media competition. U.S. policy makers must be prepared to deal with the influence of broadly based news programming that will lessen the power of American news organizations, which have been relatively supportive of American policy during recent conflicts.

Al-Jazeera

When Egyptian president Hosni Mubarak toured Al-Jazeera's cramped headquarters in Qatar, he observed, "All this trouble from a matchbox like this."[3]

For Mubarak and other Arab leaders who prefer their news media compliant, Al-Jazeera has caused plenty of trouble by fostering debate about topics that many in the region—including many news organizations—treat as being outside the news media's purview. On Al-Jazeera, everything from the role of women to the competence of governments is addressed, often loudly. The station's motto is, "The opinion, and the other opinion," which might seem commonplace in the West, but is exceptional in the Arab media world.

Al-Jazeera is the successor to a failed experiment, BBC Arabic Television (BBCATV), which was born in 1994 and quickly ran afoul of its principal financial backer, the Saudi royal family. When BBCATV aired "Death of a Princess," a documentary about the execution of a Saudi princess and her lover, the Saudis withdrew their funding and the BBC venture collapsed.[4] Some relatively moderate Arab leaders thought an editorially independent news organization might be a useful modernizing tool, and so the emir of Qatar, Sheikh Hamad bin Khalifa al-Thani, provided $140 million to hire veterans of the BBCATV experiment. They became the core staff of Al-Jazeera, which began broadcasting in 1996.

When the emir touts Qatar as a progressive Islamic state that welcomes Western investment, he showcases Al-Jazeera as evidence of his commitment to reform. He tolerates the station's independence, but Al-Jazeera's bureaus have periodically been shut down by Middle Eastern governments angered by its coverage. The station was seen mainly as a curiosity until 2001, when its content began capturing international attention. Shortly after the attacks on the United States, Libyan leader Muammar Qaddafi went on Al-Jazeera to say that he thought the attacks were "horrifying, destructive," and that the United States had the right to retaliate.[5]

Al-Jazeera also played a leading role in coverage of the U.S. war against Afghanistan. It was allowed to remain in Taliban-controlled territory after Western journalists were ordered to leave. It presented live coverage of the aftermath of American air strikes and emphasized civilian casualties and reactions to the war.[6] It gained further notoriety by broadcasting videotapes of Osama bin Laden. News organizations that were unable to get closer than the fringes of the war turned to Al-Jazeera for help, and the station's logo began appearing on newscast footage around the world.

Its constituency was growing. While it covered Afghanistan, Al-Jazeera also kept up its intensive reporting about the Israeli-Palestinian conflict, with a pro-Palestinian slant (suicide bombings were referred to as "commando operations") and emphasis on the mood on "the Arab street." Arabs in the Middle East and scattered around the world increasingly turned to Al-Jazeera.

This audience, eager for news featuring an outlook that they can identify with, is hard to define. Mohammed el-Nawawy and Adel Iskandar, authors of a book about Al-Jazeera, wrote that "the connections that bind the 300 million Arabs in twenty-two countries are often abstract. It's not a military alliance, a political truce, an economic cooperative, or a simple linguistic tie. It may not even be reduced to a common religion. Instead, what brings Arabs together is a notion of joint destiny."[7] However this population is characterized, there clearly is an audience for news presented from an Arab perspective. With that audience, Al-Jazeera has credibility that eludes Western media.

Credibility and objectivity are not the same thing, and Al-Jazeera's coverage has a pronounced tilt. Speaking about the Iraq war, Faisal Bodi, senior editor for Al-Jazeera's Web site, said: "Of all the major global networks, Al-Jazeera has been alone in proceeding from the premise that this war should be viewed as an illegal enterprise. It has broadcast the horror of the bombing campaign, the blown-out brains, the blood-spattered pavements, the screaming infants, and the corpses. . . . By reporting propaganda as fact, the mainstream media had simply mirrored the Blair/Bush fantasy that the people who have been starved by U.N. sanctions and deformed by depleted uranium since 1991 will greet them as saviors." Bodi cited Al-Jazeera as "a corrective" to the official line that Western media embraced.[8]

The approach that Bodi considers praiseworthy is viewed differently by *Newsweek*'s Jonathan Alter, who said: "Statements from Iraqi officials are covered on Al-Jazeera as facts; comments from American officials are portrayed as 'claims.' The phrase 'so-called' always proceeds 'war on terror,' and the crawl line under the screen keeps a running tally of civilian Iraqi casualties. Rumsfeld's news conference was split-screened by Al-Jazeera with a wounded girl in an Iraqi hospital bed."[9]

Choices of words and images can shape the news and the audience's perception of events. Al-Jazeera did occasionally show some restraint. While other

Arab media referred to the American-led coalition in Iraq as the "forces of aggression," Al-Jazeera used "invading forces."[10] This is not to say that Al-Jazeera pulled its punches; its coverage of the fighting—particularly its graphic depictions of casualties—fueled its critics' charges that it was sympathetic to Saddam Hussein's regime.

The station stirred considerable controversy early in the war when it showed Iraqi-supplied video of dead and captured coalition troops. Al-Jazeera's editor-in-chief Ibrahim Hilal defended these reports, stating that "what we are doing is showing the reality. We didn't invent the bodies, we didn't make them in the graphics unit. They are shots coming in from the field. This is the war. We have to show that there are people killed in this war. The viewer has to judge whether war is the most suitable way to solve problems. If I hide shots of British or American people being killed, it is misleading to the British and American audience. It is misleading to the Arab audience if they imagine that the only victims of this war are the children and women of Iraq. They have to know that there are victims from both sides."[11]

U.S. officials may have been unhappy with what they saw as Al-Jazeera's anti-American bias, but they recognized the station's clout with its more than 35 million viewers, and so set out to influence its coverage. This was part of the overall information strategy adopted by the American government, which was similar to the one that Saddam Hussein had hoped to employ—to appeal directly to the other side's public opinion and reduce the willingness to fight.[12] Using news coverage to show the other side that it could not win the war might accomplish that.

The Pentagon offered Al-Jazeera four embedded reporter slots. The station was only able to use one of these, because its reporters could not acquire the visas needed to reach U.S. units in Bahrain and Kuwait, where the governments disapprove of the channel.[13] Donald Rumsfeld, Colin Powell, and Condoleezza Rice were among the American officials who granted exclusive interviews to the station. At U.S. Central Command in Doha, Qatar, Al-Jazeera was assigned a front row seat for press briefings and its reporter was regularly called on by the briefing officer. Al-Jazeera reporter Omar al-Issawi, who was educated in Iowa and Virginia, said that "slowly, people at CENTCOM are starting to realize that we're not the enemy. We're not some insensitive monster bent on bashing America." A few days after al-Issawi said that, his colleague Tariq Ayoub was killed by an American air strike on Al-Jazeera's Baghdad office, an event that led some in the Middle East (al-Issawi not among them) to charge that the attack was a deliberate move to impede the station's coverage.[14]

Al-Jazeera's relationship with Saddam Hussein's government was also complicated. Although Saddam himself had authorized the station to cover the war, on several occasions Al-Jazeera correspondents were ordered to stop their reporting from Baghdad and leave the country. Even when those orders were rescinded (after Al-Jazeera threatened to halt all its Iraq-based war coverage), Iraqi information minister Mohammed Saeed al-Sahhaf several times publicly blasted Al-Jazeera for "taking the Coalition's side."[15] Another facet of the station's relationship with Iraq was the report shortly after the war's end that Al-Jazeera had been infiltrated by Iraqi intelligence agents. Al-Jazeera denied the charges.

During the unsettled summer of 2003, tension continued between the U.S. government and Al-Jazeera. In July, Deputy Defense Secretary Paul Wolfowitz accused Al-Jazeera of "slanting the news" in favor of Saddam Hussein, and claimed that the channel's "very biased reporting" was "inciting violence against our troops" in Iraq.[16]

In response, Al-Jazeera complained to the U.S. State Department that the channel's offices and staff in Iraq had been subjected to intimidation by American forces, including "strafing by gunfire, death threats, confiscation of news material, and multiple detentions and arrests."[17]

In September 2003, the interim Iraqi government banned Al-Jazeera (and another Arab news channel, Al-Arabiya) from government facilities and news conferences. The Iraqi Governing Council said that the stations had incited violence against the council and had fanned animosities between Shiite and Sunni Moslems.[18]

Despite such controversies, Al-Jazeera has established itself as a major media player. In addition to its large viewership, visits to its Web site increased from a million a day before 9/11 to seven million daily soon after.[19] When the Iraq war began, Internet search engines reported a surge in queries about Al-Jazeera. Lycos announced that "Al-Jazeera" had been the subject of three times more searches than "sex."[20] Hackers also targeted the Al-Jazeera site, diverting visitors to a page featuring an American flag.

Al-Jazeera emerged from the war with vastly increased name recognition and a growing audience. It remained controversial, however, within the Middle East as well as elsewhere, and that damaged its economic health. In Saudi Arabia, which constitutes 60 percent of the Persian Gulf region's advertising market, advertising on Al-Jazeera is unofficially banned because of the station's tendency to jab at Saudi officialdom. Kuwait and Bahrain have imposed similar bans. When worldwide interest in the region is high, Al-Jazeera can make money by selling footage to other news organizations, and its forthcoming English-language service may attract new ad revenues.[21] The emir of Qatar still likes Al-Jazeera as a highly visible advertisement for his country's liberalization, so he makes sure that the station can cover its $30 million annual operating costs.

Al-Jazeera's most important contributions so far may be its establishment of Arab media as a viable alternative to Western news organizations and its role in attracting global recognition of Arab media voices. As recently as the 1991 Gulf War, much of the Middle Eastern news audience had few alternatives to CNN, the BBC, and other Western media that dominated the supply of information. Al-Jazeera is now seen as their legitimate competitor.

Even with Al-Jazeera's rise, the Middle Eastern audience's hunger for news has not been satisfied. Journalistic ventures in the region continue to expand, and the interests and expectations of the news audience grow more complex.

The Evolving Arab News Media

No universally accepted definition of "the Arab world" exists, but it generally is assumed to include the 22 countries belonging to the Arab League, which have a combined population of about 280 million (roughly the same as the United States). It has the largest proportion of young people in the world—38 percent of Arabs are under age 14—and its population is expected to reach 400 million within 20 years. Although some highly visible Arabs have amassed great wealth and cartoonish stereotypes about oil-rich sheiks abound, the economy of the Arab world as a whole is stagnant. Approximately 20 percent of Arabs live on less than two dollars a day, and during the past 20 years, per person income has grown at an annual rate of 0.5 percent, which is lower than anywhere else in the world except sub-Saharan Africa.[22] Although the Arab population is centered in the Middle East, millions are scattered throughout the world.

Poverty is common and freedom is scarce. Such an environment keeps the political mood volatile, and a substantial amount of anger is perpetually simmering. Arab governments use the media to direct that anger outward—Israel and the United States are convenient villains on whom Arabs' misery can be blamed. This worldview reflects the belief that Arabs and Islam are under constant attack, and the Arab world must defend itself forcefully.

These attitudes affect perceptions of news coverage. As Arabs increasingly feel besieged by the West, they have become more skeptical about Western sources of information. The growing popularity of Al-Jazeera is just one example of Arabs turning away from Western news and relying instead on media that they can call their own.

Palestinian journalist In'am el-Obeidi wrote that during the first Intifada (1987–1993) there was ample coverage of protests against Israel. But, she recalls, the story "was always told by American or European crews addressing Western viewers, even when this reached the Palestinian audience via Arab media, such as the Jordanian or Syrian television channels." A decade later, she said, the second Intifada was covered by Arab satellite channels whose crews "were formed of locals, as familiar with the history of the conflict as they were with people's feelings and culture. Similarly, they had viewers who knew the history of the struggle, spoke the same language, and shared their feelings and beliefs. For the first time, Palestinians felt that they were no longer subjects of an outside narrator. They felt that their story was being told and narrated by themselves."[23]

The number of Arab news outlets, such as satellite television channels, is growing. Besides Al-Jazeera, the television choices include Abu Dhabi, Al-Arabiya, LBC (Lebanon Broadcasting Corporation), Arab News Network, Al-Alem, Al-Manar, and others. In print, the range of choices is far greater, with "quality" Arabic newspapers, some edited in London and elsewhere outside the region, facing competition from local publications large and small. Rami Khouri, executive editor of Beirut's *Daily Star,* said that "like their audience, the Arab world's newspapers are angry, nuanced, multifaceted, passionate, and argumentative." He noted that the increased availability of satellite television, FM radio, and on-line news has made the news business even more competitive, giving the audience the opportunity to compare the information they are receiving from different sources. "Arabs," he said, "are increasingly tired of being lied to and presented with only half of reality, and their press is starting to reflect this."[24]

Unlike much of the Western audience, Arab news consumers are able to comparison shop for news because they have access to American and other Western media as well as their own. Among the Arab media, satellite news channels—with their pan-Arab approach—were preferred over the more parochial national television stations during the Iraq war. This pushed stations such as Egypt's Nile News to adopt a sharper-edged journalism in order to compete with Al-Jazeera and other pan-Arab news stations. The satellite channels were being seen not just by those who could afford their own dishes, but also by large numbers of Arabs from all levels of society who gathered in neighborhood coffee shops to watch. Newscasts and talk shows almost uniformly depicted the United States as an invader, not a liberator, and addressed the "clash of civilizations" and various conspiracy theories (often involving a U.S.-Israel plot) about the war.[25]

Coverage of the Iraq war was shaped partly by the larger context in which some Arab news organizations placed it. On one occasion, the Saudi newspaper

Al-Watan fused photographs of suffering Arab civilians from the Palestinian territories and images from Iraq to make the point that the United States and its allies were engaged in a long-term assault on Arabs.[26] Beirut-based Al-Manar television, operated by Hezbollah, makes clear in its statement of purpose that its coverage conforms to a political agenda: "Political, cultural, and social affairs are of special importance to the station's programs. Most important is the struggle with the Zionist enemy."[27]

For many Arab television stations, part of the wartime agenda was to compete effectively against Al-Jazeera. Abu Dhabi TV, based in the United Arab Emirates, adopted an all-news, 24/7 format when the war began. It deployed ten satellite news gathering trucks and equipped its correspondents with satellite videophones so they could report live at any time.[28]

Another newcomer was Al-Arabiya, a satellite news station that began its around-the-clock all-news operations in early March 2003. It is a Dubai-based affiliate of Middle East News and the Middle East Broadcasting Center, which is owned by a brother-in-law of Saudi King Fahd.[29] It billed itself as a "balanced alternative" to Al-Jazeera that would avoid "deliberate provocation" in its coverage.[30] Al-Arabiya quickly encountered problems. Because it was so new, it did not receive accreditation from the coalition's Central Command, and so could not attend briefings in Doha. Its crew covering the fighting in Iraq was captured by Iraqi forces, but was released when a tribal chieftain decided he did not want be a jailer for Saddam's collapsing regime.[31] With those problems finally out of the way, Al-Arabiya faced the longer term task of trying to win audience with its "balanced" approach, which may be journalistically sound but might not attract viewers who like Al-Jazeera's uninhibited talk shows and other lively news programs. In November 2003, the Iraqi Governing Council banned Al-Arabiya from broadcasting from Iraq after the station carried a taped message purportedly from Saddam Hussein that called for attacks on Iraqis who cooperated with American occupation forces.

By this time, Al-Arabiya, along with Al-Jazeera, had established itself as a favorite of Iraqi viewers. Satellite dishes had quickly become a hot commodity in markets in Baghdad and elsewhere, and by late 2003 about one-third of Iraqis had access to satellite broadcasts. A poll conducted for the U.S. State Department in October 2003 found that among Iraqis with this access 63 percent got their news from Al-Jazeera and Al-Arabiya, while 12 percent relied on the Iraqi Media Network, which operates under a contract from the U.S. Department of Defense.[32]

The Iraq war gave Arab news media an opportunity to engage in critical reporting and to cover events in the Arab world in the broader context of global politics. Questions remained, however, about whether this marked a lasting change from the defensive outlook that had long kept many non-Arab viewpoints out of the news.[33] *Washington Post* columnist David Ignatius, speaking at the invitation of Hezbollah to a post-Iraq war conference, told the audience, "The only thing that worries me about the rise of the Arab media is that they sometimes see their job as telling the story from the Arab point of view—rather than just telling the story. . . . The Arab people deserve to know the truth, even when it hurts."[34]

While Arab news organizations searched for the political path that they wanted to follow, news continued to flow, sometimes from more traditional sources. Associated Press Television News (APTN) offered "Middle East Custom Coverage" to satellite and terrestrial stations, and the organization's CEO, Ian Ritchie, said that despite the temptation to avoid APTN as being too Western, "I cannot think of a single Middle East broadcaster" that wasn't using the APTN service.[35] Similarly, other Western news organizations, particularly the BBC, retained a following among news professionals and news audiences in the region.

Nevertheless, the dominant story line from the region's media was evil America versus noble Iraq. This did not, however, find universal acceptance in the region. Qatari writer Abdulhamid Al-Ansary asked, "Why did the Arab media consent to align itself with the Iraqi regime while at the same time pretending it was with the people?" The answer he heard from Arab media executives was that the news was shaped to fit the demands of competition for the prime audience—"the Arab street." "The street," said Al-Ansary, "is emotional and has little confidence in the Americans. It can be won by fanning the flames of its emotions and encouraging its feelings with dreams of a great Arab victory and a great American defeat."[36]

With emotions high, television coverage had particular impact on the public. Writing about Egyptians' response, Hussein Amin explained that "the image of an American flag draped over Saddam Hussein's statue was transmitted to tens of millions of Arab viewers and contributed to a sense of the humiliation of their Arab brothers and their fears of American imperialism. This is an excellent example of the power of transnational satellite broadcasts—one soldier makes an individual gesture and an entire region watches in astonishment."[37]

Some Palestinian college students noted that a CNN correspondent talked of "disturbing" news when he reported about American casualties, but used no such modifier when describing Iraqi casualties.[38] Nihal Saad, a correspondent at Nile TV International said: "The world is watching two entirely different wars.

Everybody is reporting from their perspective of the truth. We of course think we are accurate." And accuracy, as perceived by one Egyptian farmer, was that "America has killed thousands of Iraqi children. They want to destroy Islam as a religion."[39] His opinion was far from unique.

When Baghdad fell, much of the Middle East coverage was subdued. Syria's official newspapers made no reference to the American takeover of the city, and papers such as Egypt's *Al-Akhbar* used headlines such as, "Surprising Collapse of the Saddam Regime."[40] Lebanon's LBC channel reported that some Iraqis were welcoming American troops, but such stories were rare,[41] as were editorials such as the one in the Saudi newspaper *Al-Youm* that reflected some of the previously suppressed dislike of Saddam, saying that the toppling of Saddam's statue "expressed the removal of a nightmare that has been lying on the chests of the Iraqi people." The editorial went on to distinguish between Saddam and most Iraqis, stating that "The only thing that can be said is that this is the fate of any dictator who murders his people, treats them brutally, and steals their resources. Had the people supported Saddam, no force, no matter how strong, could have conquered Iraq."[42]

Some of the coverage looked at what lay ahead for the Arab world and the United States. The Palestinian newspaper *Al-Quds* editorialized that "This catastrophe, which adds to the series of Arab defeats throughout the last century, may create a new consciousness and profound clarity of mind within the Arab nation and Muslim world, since everyone knows they are in a state of regression, cultural backwardness, and disintegration, being unable to cooperate with each other and with the entire world." Another Palestinian paper, *Al-Ayyam*, warned Americans that they "are paving a long, broad path for the death of tens of thousands, maybe even more, of their people. The American madness will bring nothing but counter-madness. They have begun an era of destructive and lethal war in order to feed their aggressive military-economic machine, and they will bear the responsibility for it."[43]

Rami Khouri, of Beirut's *Daily Star*, stepped back from the emotions of the moment and looked at the long-term political ramifications of the American victory. "The sacking of Baghdad," he said, "is designed to send signals to all other Middle Eastern and Asian regimes that the U.S. finds annoying, threatening, distasteful, worrisome, or even just a little strange. . . . If Washington merely suspects that terrorists may one day emerge from your land, or that you might in future threaten your neighbors, you have only two options: You change course and shape up, or you are finished as a governing regime. If you behave as Baghdad behaved, defying the new rules of the game, you suffer the same fate as Baghdad is suffering."[44]

The American assertiveness that Khouri addressed was a major story of the war. Around the world, it was receiving at least as much news coverage as was the fighting itself.

The Larger World: News Coverage Elsewhere

Just up the road from the Arab world are Turkey and pockets of the region's Kurdish population—important constituencies with considerable interest in Iraq's future. Their media reflect a mix of nationalist programming and material from Western news organizations.

During the 1991 Gulf War, Turkey's first private television channel, Star TV, delivered CNN to its viewers, as did Turkish Radio and Television (TRT), Turkey's public broadcasting company. By 2003, there were 16 national, 15 regional, and more than 220 local channels in Turkey, including all-news CNN-Turk and NTV, which is a partner of MSNBC. CNN-Turk and NTV competed fiercely, advertising their resources for covering the fighting in Iraq.[45]

The array of media is large, with some Western influence—as with those two news channels—but also with many truly indigenous ventures. In addition to the television offerings, there were at least 21 national daily newspapers and more than 3,000 local papers (some published daily, some occasionally), plus more than 100 magazines to serve Turkey's population of more than 67 million.[46]

Despite the use of CNN and MSNBC material, much of the television coverage during the Iraq war emphasized Turkey's perspective on events, particularly the strained relations between Turkey and the United States caused by Turkey's refusal to allow American troops to invade Iraq from Turkish territory. Several major stations used retired Turkish military officers as commentators and the tone of much of the coverage was deferential toward the United States, with commentators discussing prospects for restoring good relations between the two countries. Other stations, especially those with a more strongly Islamist orientation, expressed more concern about Iraqi civilians and less about Turkish-U.S. affairs.[47]

This sketch of change in Turkey illustrates how quickly and thoroughly a country can be brought into the mainstream of international communications. During the decade between the two U.S.-Iraq wars, the flow of news from external sources into Turkey increased dramatically, largely because of satellite television. The presence of international media organizations such as CNN and MSNBC spurred intensified competition among Turkish media companies. The Turkish government found that the number of voices, some of them critical and far beyond the reach of any government-imposed control, had multiplied, changing the political dynamics of the country.

Meanwhile, Turkey's more than ten million Kurds—always a source of concern to the central government—could watch the war on Turkish stations and on Medya, a new Paris-based satellite channel that reaches Kurdish-language viewers in the Middle East, North Africa, and Europe. Medya's mission statement promises to "ensure that Kurdish-speaking audiences are provided with a full array of television programs which reflect the broad needs of the Kurdish people. . . ."[48] That is a political as well as cultural promise to the Kurds, who have a nationalist agenda that is often seen as running contrary to the interests of the countries in which they are living. Medya helps bring coherence to the concept of Kurdish nationhood.

Just as the war intensified the interest of many Turks in international news, coverage of the conflict also saw changes in news preferences in countries that had long had access to international sources of coverage. The BBC won new viewers, often at the expense of American news organizations such as CNN that were perceived as being biased.

The Turkish audience's move away from reliance on U.S. news media was not unique. During the war, BBC World's ratings increased 28 percent in the United States, with its programs being watched in 662,000 American households. Around the globe, BBC World reaches 254 million homes in 200 countries. BBC radio broadcasts are translated into 43 languages and reach 150 million listeners, and BBC's online news, also in 43 languages, attracted 220 million page views (58 million from the United States) during March 2003.[49] On just the first day of the war, the BBC News Web site had 33 million page views and received more than 27,000 e-mails.[50] This surge took place while "the Beeb" was drawing political fire from both the left and the right (usually a good sign). It was called "the Blair-Bush Corporation" by some and "the Baghdad Broadcasting Corporation" by others.

The BBC's audience gains during the war should serve as notice to American news executives that they cannot take even the U.S. audience for granted. If news content has a chauvinistic tilt, some members of the audience will look elsewhere. That said, it is unrealistic to think that all news coverage will be wholly objective and not colored to some degree by politics, particularly when political leaders are intent on stirring up public emotion.

This could be seen in the German headlines mocking the slowdown in the American advance on Baghdad (*Der Spiegel:* "Superpower in the Sand: America's Stuck Blitzkrieg") and focusing on Iraqi civilian casualties (Berlin's *Der Tagesspiegel:* "Bloodbath in Baghdad Market").[51]

The German public followed the war closely, as evidenced by the unusually high ratings for the country's two all-news channels. In Germany, as elsewhere, news about the 1991 Gulf war came primarily from CNN, either directly or as picked up by German broadcasters. There were no German all-news channels then. By 2003, Germans could watch N-TV, which was owned by AOL Time Warner and Bertelsman and had a CNN look, and N24, another privately owned channel. These news stations have so far captured only a small audience, but during the first week of the war N-TV's small share of the German television market more than doubled, up to 2 percent, and N24's share nearly tripled, to slightly more than 1 percent. N-TV and N24 produce their own coverage as well as using material from CNN, Reuters, Al-Jazeera, CBS, and other sources. Their war coverage differed from that of their American counterparts, placing more emphasis on civilian casualties and more frequently featuring Al-Jazeera's reports. Another news channel, tax-supported Phoenix, offered a C-SPAN approach, providing uninterrupted coverage of Security Council debates, Pentagon briefings, German political events, and such.[52]

Coverage of the war in much of the rest of the world had a pronounced anti-American tone. Even in Spain, where the government was supporting the Bush administration, top newspapers editorialized against the war and against Spain's endorsement of it. *El Pais* said: "Although the scent of victory distracts those who carried it out, this war was avoidable. The world is better without this dictator, but the management of this conflict contributes to debilitation of the already fragile international order."[53]

In France, *Le Monde* criticized U.S. military tactics as being "a flood of fire versus the slightest threat or what is perceived as such: air raids and tank fire and heavy machine gun fire in a crowded downtown. The civilian victims undoubtedly number in the hundreds. A military culture is the cause: the massive use of force against the slightest danger, so much the worse for civilians."[54] Mexico's *El Universal* challenged American motives: "The Iraqis at least know the reason they are dying. Not so the American soldiers. The Iraqis defend their mother country, not the tyrant Saddam Hussein. The American soldiers, many unwittingly, defend the economic interests of the most powerful corporations, directed by the leaders of their government and army, all representatives and beneficiaries of the American military-industrial complex."

In Kenya, the *Daily Nation* criticized American news coverage: "What is dreadful is that during 'peace' these same media bombard the world with holier-than-thou sermons about objectivity, truth, and fairness, whereas in war they are the first to trample these principles underfoot." North Korea's *Nodong Sinmun* saw the Iraq war as a possible prelude to an attack on its own country.

It claimed that "the United States imperialists' provocation of a war in Iraq shows that they, who are the world's worst rogues who are blindly beating a legal sovereign state while ignoring both the United Nations and international law, can wage a greater war of aggression against our country at any time."[55]

Russian media coverage, like the Russian government's stance on the war, was more complex than that of many other countries. When the U.S. advance on Baghdad had apparently stalled, Russian television reports compared the situation to the stalemate in Chechnya, implicitly telling viewers that "We're not the only ones who can get stuck; even the all-powerful American military has problems."[56] *Izvestia* acknowledged that Russia's unwillingness to join the U.S. war effort was partly due to Russian business interests in Iraq and the eight billion dollars the Iraqi government owed Russia. The newspaper also criticized some of the Russian television coverage as being too critical of Bush and the United States, arguing that "it is one matter to simply refrain from supporting the war against Iraq and quite another matter to become a sworn enemy of America." The editorial recognized a shared interest in fighting terrorism: "Being at odds with America is stupid, but not because America is stronger than Russia. There is a more weighty reason: we have a common enemy and common ideals. Russia can and should call on the U.S. to end the bloodshed in Iraq as soon as possible. But that call should be directed to a companion-in-arms in a common cause who is making a mistake, not to an enemy."[57]

That passage says something about the Russian political psyche. Russia was defensive about its struggle with Chechnya and wanted America's assertive action against Iraq to be seen as comparable to that conflict, with both Russia and the United States as righteous warriors fighting terrorism. "Not because America is stronger than Russia" sounds plaintive, but it reflects the mood of a superpower that is no longer super. Russian president Vladimir Putin knew he needed the United States over the long term as a political and economic partner, but it was difficult to embrace a nation that was the principal adversary for so long. As it has done for many years, *Izvestia* was sending signals as well as delivering news.

Another interesting place for examining war coverage was Indonesia, the world's largest Muslim country. CNN and Al-Jazeera are both readily available there. During the war, in addition to directly reaching viewers with its own signal, Al-Jazeera was carried by Indonesian network TV–7 for 12-hour stretches, with local interpreters translating the Arabic. Two other Indonesian networks also carried Al-Jazeera in shorter segments.

CNN and Al-Jazeera often presented conflicting reports, particularly about the degree of success of the coalition forces and the impact of the fighting on Iraqi civilians. Reporting from Jakarta, the *Washington Post*'s Ellen Nakashima found public suspicion of CNN, with some Indonesians seeing it as a voice of the American government. CNN and other Western news organizations might proclaim their independence, but in much of the world, where "free press" has long been a problematic concept, skepticism is hard to dispel. Words and images presented by Al-Jazeera were seen as more credible because of the channel's Islamic credentials. Nakashima also found that war coverage in many Indonesian newspapers had an anti-American tone, partly, said one journalist, for commercial reasons—that was the kind of news product the public wanted to buy.[58]

Throughout all the coverage from around the world, overt support for Saddam Hussein was virtually nonexistent. Nevertheless, there was overwhelming suspicion of American intentions, with little visible backing for the idea that it might be worthwhile to get rid of a tyrant who had brutalized his own people and had helped destabilize an entire region. In the eyes of many of the world's news organizations, and presumably many of the world's people, the United States almost always operates under a presumption of guilt.

Stereotyping, even among putative friends, casts a shadow on relations among governments. Among the kinder characterizations of George Bush was that of six-gun-wielding cowboy; he was also portrayed as a latter-day Hitler. Britain's *Daily Mail* described his policy as "insane warmongering." Numerous critics said that the real reason the United States went to war was to acquire Iraq's oil riches.

While receiving so much criticism, Americans also dished out insults. France became a favorite target, and the collateral damage included Belgian-born French fries, unappetizingly rechristened "freedom fries."

On a more substantive level, former *Financial Times* editor Richard Lambert observed that the "idea that Europe is declining to the point of irrelevance in military, economic, and political terms frequently recurs in U.S. commentaries." The public was provided with little journalistic context for evaluating these commentaries, as news coverage often featured simplistic shorthand. Lambert analyzed the use of country names and related topics in the top 20 U.S. newspapers in 2002. He found that "Germany" and "Schroder" appeared together less often than did "Germany" and "Hitler." When he looked at "French," he found that "cooking" was the linked word more often than "politics" or "economics."[59]

That reflects American news organizations' generally superficial approach to international coverage, and the ignorance such flimsy coverage leaves in its wake affects the politics of American foreign policy. When Donald Rumsfeld defines relationships between the United States and European nations in terms of "old" and "new" Europe, he is making an interesting point, and one that merits public debate. But Americans are ill-prepared to engage in debate about revising alliances or to consider the ramifications of a go-it-alone approach in dealing with the rest of the world. They watch their intellectually mushy news coverage, chomp on their freedom fries, and let the world pass them by.

The roots of criticism of U.S. policy toward Iraq spread in many directions. Much of that criticism was thoughtful and deserved equally thoughtful consideration even from those who disagreed with it. Writing in Germany's *Suddeutsche Zeitung*, Helmut Schaefer said: "All those who brand our reticence as a lack of solidarity, as ingratitude, or even as pacifism should remember that we owe a great debt to the U.S. for contributing to our transformation as truly democratic citizens after World War II and Hitler's dictatorship. They must forgive us if we have difficulty letting go some of the lessons we have learned." There are also vast differences in perceptions of the struggle between Israel and the Palestinians, and not in the Middle East alone. Much of the world is not as supportive of Israel as the United States is, a fact lost on most Americans. Yet another important factor shaping global outlooks is the lasting impact of the 9/11 attacks. The events of that day will be a driving force in U.S. policy for years to come. Many throughout the world were sympathetic at the time, but that feeling faded. Most other governments, correctly or not, do not act as if their own nations are at risk from major terrorist attacks.[60]

When it became clear that the American invasion of Iraq was going to succeed, some of the critics began to reassess their relationships with the United States. In Germany in early April, *Die Welt* urged the government to develop a foreign policy "that realizes that neither the European Union, NATO, nor the United Nations can counterbalance America."[61]

This was an example of politically expedient America-bashing giving way to pragmatism. Popular or not, the United States—particularly a triumphant United States—could not be ignored. Journalistic and political perceptions are closely related. Like much of the political rhetoric around the world, much of the international news coverage before the war and during the early stages of the fighting featured anti-war and anti-American viewpoints. After the capture of Baghdad, politicians and the news media began adjusting their outlooks to conform to a new reality in which America, regardless of international opinion, was certain to play the dominant role.

Iraq Seeks Its Voice

When Iraqis themselves emerged from the war they tried to bring a liberated coherence to many aspects of their national life, including the news media. Anthony Borden, executive director of the Institute for War and Peace Reporting, said that "the challenge in Iraq is to create a publicly owned media system, moderate in tone and dedicated to responsible reporting, that would reflect the country's diversity of views and competing interests. The concept may clash with the U.S. instinct for private institutions. But it is a useful halfway house in moving firmly away from Iraq's state-controlled media, dominated by a Ministry of Information and infiltrated by the security services." Borden warned that highly partisan media organizations might reflect the country's new freedom, but in a transitional system could foster a "vendetta journalism" that might provoke violence and impede national reconstruction. He urged that the BBC be used as a model, with a board of governors that would set guidelines for operations and content.[62]

Iraqi media organizations faced a steep climb. Before the war, Uday Hussein, one of Saddam Hussein's murderous sons, was in charge of the Union of Journalists, and his control smothered press freedom. Reporters and editors knew that certain words, such as "democracy," could not be used and institutions such as the oil industry could not be examined.[63] Millions of Iraqis had fled the country to escape Saddam's regime, and some of them established an Iraqi exile press in cities such as London and tried to deliver news to their homeland.

Some of these journalists returned after the war, ready to be media entrepreneurs in the new Iraq. Saad al-Bazzaz came home from London, where he had been editing *Al-Azzaman*. He began publishing editions of the paper in Basra and Baghdad, mixing political stories and reader-attracting fare such as large pictures of Jennifer Lopez. Because of the lack of printing facilities in Basra, he had that city's edition printed in Bahrain and trucked into Iraq. For his Baghdad operation, he imported computer and satellite equipment and provided content generated in London to a local printer. He began with a press run of 30,000 copies and reported that vendors were selling them all.[64]

Iran also took advantage of the opportunity to deliver its version of news to Iraq. Al-Alam television, presumably supervised by the Iranian government, broadcast from just across the border, close enough to Baghdad so its signal could be received with a regular antenna; no satellite dish was needed. Using Al-Jazeera's coverage as well as its own reports, Al-Alam's smoothly produced programming was available during and after the war. American officials watched it warily, wondering if the Iranian government would use it to stir up trouble, particularly among Iraq's Shia population.[65]

Most of the new media enterprises appeared to be truly local. In central and southern Iraq, about 150 new publications appeared within weeks of the war's end. Some had clear political agendas, such as *Al-Saah*, published by a Sunni cleric who had a religious program on Abu Dhabi television. The Shia paper *Sadr* took an anti-occupation line and called for Iraq to become an Islamic republic. In addition to the indigenous products, two papers were funded by the U.S. government and produced in Kuwait for Iraqi readers.

Although much of the media activity was centered in Baghdad, new print and broadcast products appeared elsewhere in the country as well. In Basra, news programming on Radio Nahreen was produced by a British Army Psychological Operations unit. Looking for an eventual revenue stream, the station gave free advertising to restaurants and other local businesses in return for their placing advertisements for the station on their premises. In Najaf, Najaf Television resisted efforts to be brought under the control of the local American administrators and provided Islamist-oriented programming to a small area.

Most of these outlets were in the Kurdish regions that had been outside Saddam's control since 1991. Between Erbil and Suleimaniya, there were 242 print and 102 TV and radio stations reported to be in operation. Many of these were affiliated with one or the other of the two leading Kurdish political parties. After the war ended, many of the Kurdish media focused on representing Kurdish interests in anticipation of dealing with a new central government.[66]

Lots of activity was going on, but there were also many fundamental problems. The majority of experienced journalists had political ties to the Saddam Hussein government, the Baath Party, or opposition groups. Political agendas can be difficult to set aside, and so can distort coverage. Worse, the news media could be tools used to foment disorder. In late May, American occupation administrator L. Paul Bremer issued rules to govern operation of Iraqi media, including the requirement that all media operations be registered. The rules banned "hate speech" that might trigger violence, support for the Baath Party, and material that was "patently false" and designed to provoke opposition to the occupation forces.

American officials shut down a twice-weekly Baghdad newspaper, *Al-Mustaqilla*, after it ran the headline, "Death to All Spies and Those Who Cooperated with the U.S.; Killing Them Is a Religious Duty." A coalition spokesperson said: "This is not about censorship. This was too dangerous, a flagrant breach of international law. This was a newspaper that was encouraging violent attacks and encouraging the killing of Iraqi nationals. It was our duty to close down this newspaper." Although it was clear that *Al-Mustaqilla* had gone too far, some Iraqi journalists protested its closing. One reporter at *Al-Zaman*, a major Baghdad newspaper, said: "This is a violation of press freedom. Freedom means having opinions, even clashing ones."[67] The rules of public debate had clearly changed. Anyone protesting a violation of press freedom just a few months before would not have survived for long in Baghdad.

In an attempt to improve stability, U.S. planners set up the Iraqi Media Network (IMN) as the officially sanctioned venture to produce print, television, and radio news. Its efforts were criticized for being overly bureaucratic, insufficiently appealing to the mass Iraqi audience, and generally inept. A study of the Iraq media situation for two British and two Danish NGOs found that the IMN "lost a critical opportunity to present a fresh and dynamic face for a new Iraq, and failed in its duty to provide basic information to a frightened and traumatized population." The report added that even with the "explosion of open speech, the provision of balanced reporting, especially about local issues and humanitarian concerns, remains scant."[68]

Improving the quality of reporting over the long term would require training a new corps of Iraqi journalists, with particular attention to coverage of issues that would be crucial in resurrecting Iraq, such as human rights, cross-cultural cooperation, and the economic and political dynamics of nation building.

———

People throughout much of the world had access to more information about the Iraq war than about any previous conflict. That was largely a function of communications technology that enhanced production and delivery of everything from cheap newspapers to satellite television to Web-based news bulletins. The quality of news was a different matter altogether. It has gotten better, but not at the same pace as the mechanics of the news business have advanced.

Policy makers must deal with the fact that more of the world is getting more information than ever before. In years past, the American government could count on being able to deliver its message to the world through American news organizations, particularly those using satellite transmission. That made CNN an invaluable policy tool. The first Gulf war was the high point for this method of influencing world opinion.

Little more than a decade later, the balance of information power had changed drastically. During the Iraq war, American media voices no longer held the world's attention by default, and those who made the case for American policy encountered opposition that was loud, persistent, and far-reaching.

In the future, when the United States makes its case for its policies—sometimes including military action—it is certain to encounter pervasive and potent opposition. To deal with this and to give U.S. policy makers a better chance of winning the battle for world opinion, the arts of diplomacy must adapt more fully to the demands of a media-centric world.

CHAPTER SEVEN

Media Diplomacy:

Escalating the Battle for Hearts and Minds

A superpower at war must decide how much effort it wants to put into justifying its actions. If the outcome of the conflict is certain, the temptation exists to just fight, win, and be done with it. But when the superpower also cares about what people around the world think of it, then it must make its case. In doing so, news coverage can be useful, but other media tools must also be brought into play. "The media" consist of more than traditional journalism venues.

In the aftermath of the 2001 attacks on the United States, American policy makers were particularly sensitive about how the country was portrayed. "Why do they hate us?" is difficult to answer when political leaders resist introspection and resent having their intentions challenged. Although the Bush administration grudgingly accepted the need to be seen as something other than a selfish bully, the negative image of America became more fixed the farther the events of 9/11 receded in time. In 2003, the U.S. war against Iraq was widely viewed as unjustified, generating more of the hatred that seemed to so mystify American policy makers and public.

In response, the White House defined the invasion of Iraq as a war of liberation and produced a media campaign to support that idea. The official theme was "Iraq: From Fear to Freedom," and the news media were given "liberation updates" and heard "voices of freedom" from Iraqis who appreciated Saddam Hussein's ouster. The administration's product was well planned and slick, but it had to contend with reality, or at least the versions of reality that much of the world was seeing. For every picture of American soldiers providing medical care or delivering food, there were more images of dead Iraqi civilians.

One Arab journalist said the public in the Middle East saw the American actions as a "feed-and-kill" policy, and Egyptian president Hosni Mubarak warned that television coverage of civilian casualties was creating more Osama bin Ladens.[1]

Complicating matters were false and inflammatory stories that some news organizations, especially in the Middle East, were running. American officials mounted a "rumor patrol" to knock down reports such as those that accused the United States of targeting mosques in air strikes.[2]

Given the number of media messages that people throughout the world were receiving, and considering how many of those messages were coming from sources unfriendly to the United States, American officials realized that their own information efforts had to be taken to a new level. "Propaganda," with all its pejorative connotations, needed to be superseded by "public diplomacy."

Public Diplomacy as Policy Tool

A baseline definition of public diplomacy is to inform, engage, and influence foreign publics.[3] A lengthier explanation, offered by diplomat Christopher Ross, is that "the meat and potatoes of public diplomacy is giving timely news to foreign journalists, providing information on America directly to foreign publics through pamphlets and books, sponsoring scholarships and exchanges to the United States, exhibiting American art, broadcasting about U.S. values and policies in various languages, and simply transmitting balanced, independent news to captive people who have no information source independent of a repressive government."[4]

The principal difference between this wide-ranging mandate and traditional diplomacy is that the latter is based on a government-to-government relationship, while public diplomacy builds links between a government and foreign publics. Joseph Duffey, former director of the United States Information Agency, said that public diplomacy is "an attempt to get over the heads or around diplomats and official spokesmen of countries and sometimes around the press to speak directly to the public in other countries and to provide an interpretation, explanation of U.S. values and policies."[5]

Duffey's point about getting around the press underscores the fact that public diplomacy is an information medium that in some ways is in competition with the news media. U.S. policy makers recognize the importance of news coverage in telling the world about American actions and principles, but they quickly run into a significant problem: American news organizations pride themselves (correctly or not) on remaining objective, rather than purposely supporting government policy. Because they do not provide a consistently partisan view of U.S. policy, and because many news organizations from elsewhere in the world are overtly anti-American, the U.S. government needs an advocate. That is where public diplomacy comes in.

Although public diplomacy has been in vogue since 9/11, it has its critics. *The Washington Post*'s Jim Hoagland wrote: "Diplomacy is not public relations, a crucial point this administration has yet to demonstrate it understands. Nor is diplomacy politics. Diplomacy involves the calculated use of national leverage where it exists—and of persuasion and camouflage where it does not—to get other governments to fit their policies to yours. Even in this electronic era, the task of diplomacy is to move governments, not outflank or upstage them by trying to change the hearts and minds of their citizens overnight."[6]

Public diplomacy exists to bring a new dimension to traditional international relationships. For the United States, the need for such an effort is clear. Testifying before Congress in February 2003, Andrew Kohut, director of the Pew Research Center for the People and the Press, said, "The most serious problem facing the U.S. abroad is its very poor public image in the Muslim world, especially in the Middle East/Conflict Area." Citing results from a survey of 38,000 people in 44 countries, Kohut noted that in Egypt only 6 percent of the people held a favorable view of the United States, and in Lebanon, Pakistan, and Jordan, more than half the people surveyed had a very unfavorable view of the United States. Similar results came from NATO member Turkey, where only 30 percent had favorable feelings about America.

In that region, the dislike centers not on America in a general sense, but on U.S. policy, particularly America's support for Israel and indifference toward Palestinians. More broadly, people in Muslim countries tend to see the American war on terror as a war on Islam.

Kohut reported, however, that "it is not all one way—even in Muslim countries, opinions about the U.S. are complicated and contradictory. As among other people around the world, U.S. global influence is simultaneously embraced and rejected by Muslim publics." American technology and popular culture find approval even in some of the countries with strong overall anti-American attitudes. Nevertheless, in a broader sense, the spread of "Americanism" attracts criticism.

The most encouraging news to emerge from the survey, said Kohut, was "a very substantial level of democratic aspirations among Muslim people. People in Muslim countries place a high value on freedom of expression, multi-party systems, equal treatment under the law—in fact, higher than in some nations of eastern Europe." This means that America might recoup some of its lost support by emphasizing policies that encourage democratization.[7]

Kohut's testimony was not the first warning about such problems. Concerning the media coverage that U.S. officials found troubling, Middle East expert Shibley Telhami had pointed out that "in large part, Al-Jazeera's success springs from its ability to reflect public opinion, not to shape it."[8] Graham Fuller, former vice chairman of the CIA's National Intelligence Council, cited Arabs who say: "We want your political values. It is you we perceive as not applying them in any consistent way." Fuller also said, "Clearly, in a region where we desperately need friends and supporters, their number is dwindling, and we are increasingly on the defensive." Rep. Henry Hyde, chairman of the House International Relations Committee, reflected widespread frustration when he asked: "How has this state of affairs come about? How is it that the country that invented Hollywood and Madison Avenue has allowed such a destructive and parodied image of itself to become the intellectual coin of the realm overseas?"[9]

At issue is not simply wanting to be popular. A Council on Foreign Relations task force in 2003 found that "anti-Americanism is endangering our national security and compromising the effectiveness of our diplomacy. Not only is the United States at increased risk of direct attack from those who hate it most, but it is also becoming more difficult for America to realize its long-term aspirations as it loses friends and influence. By standing so powerful and alone, the United States becomes a lightning rod for the world's fears and resentment of modernity, inequality, secularism, and globalization."[10]

This report also strongly criticized the American response to the situation, contending that the United States "has significantly underperformed in its efforts to capture the hearts and minds of foreign publics. The marginalization of public diplomacy has left a legacy of underfunded and uncoordinated efforts. Lack of political will and the absence of an overall strategy have rendered past public diplomacy programs virtually impotent in today's increasingly crowded communications world."[11]

America's public opinion problems are by no means limited to the Middle East. In Europe, for instance, polling conducted in June 2003 found that 70 percent of respondents in France, 50 percent in Germany, and 50 percent in Italy believed that global U.S. leadership is "undesirable." The survey also found that Germans, in particular, saw the European Union as being more important to their vital interests than the United States is. When European opinions were compared to those of Americans, the Americans were consistently more bellicose in their responses about dealing with the Middle East, North Korea, and Iran.[12]

Given these and similar findings, few would dispute America's need to improve relationships around the world, but how to do so always inspires debate.

Traditionally, such efforts had been the primary responsibility of the United States Information Agency (USIA), which was created by President Dwight Eisenhower in 1953 as an executive branch agency with a mandate to counter Soviet-generated anti-American propaganda. Its directors included journalists such as Edward R. Murrow and Carl Rowan, and among its broadcasting ventures, the agency operated the Voice of America radio service and WORLD-NET, a satellite television channel, which along with other broadcasting operations were consolidated within the International Broadcasting Bureau in 1994.

After the Cold War ended, USIA became a target for Congressional budget-cutters, and in 1999 it ceased to exist as an independent agency. It was rolled into the State Department, where it had a much lower profile.

Predictably, being absorbed into a larger bureaucracy meant reduced dynamism and fewer accomplishments. Information-related efforts were less cohesive and less effective. Public opinion needs nurturing; without it, the public's mood drifts this way and that, and negative voices have disproportionate effect, particularly if unanswered. This was the state of affairs that the Bush administration faced in the aftermath of the 2001 attacks. Veteran advertising executive Charlotte Beers was named Undersecretary of State for Public Diplomacy and Public Affairs, and the White House, with efforts led by presidential counselor Karen Hughes, created its own public diplomacy operation. The Coalition Information Center coordinated the post-9/11 information flow and worked with British and Pakistani officials during the fighting in Afghanistan.

This ad hoc project gave way in 2002 to the Office of Global Communications, partly because the administration recognized that the United States had fared badly in the coverage of Afghanistan by Arab media. During the pursuit of the Taliban and Al-Qaeda, the United States had to deal at length for the first time with Al-Jazeera and other high-tech and high-energy Arab news organizations. American officials were unprepared for the tough tone of the reporting and had underestimated the reach and impact of this coverage. So, before the fighting in Iraq began, the White House designed its new in-house operation to meet the demands of the global 24-hour news cycle. A daily conference call among the media relations chiefs at the White House, State Department, Pentagon, and 10 Downing Street determined what story lines to emphasize and what problems to address. Around midday U.S. time, the Central Command in Qatar held its daily briefing, and a few hours later Department of Defense officials conducted their own session at the Pentagon. Throughout the day, administration officials were scheduled for interviews with American and international media. At day's end, the White House office prepared the "Global

Messenger," with talking points and quotes from the president and others. This was e-mailed to Washington officials, U.S. diplomats around the world, and other people who might be in contact with journalists.[13]

The whole process was reminiscent of a well-disciplined "war room" in a political campaign. Although the stakes were far higher in this real war, this was a sensible model to employ in keeping the U.S. government, like a political candidate, "on message."

The efficiency of this operation could be seen when administration officials told the same anecdotes to justify policy decisions. A problem arose, however, when the accuracy of some of those anecdotes was questioned. One story that was told by President Bush, Secretary Rumsfeld, Pentagon briefers, and others was that an Iraqi man who had criticized Saddam Hussein was tied to a post in downtown Baghdad, his tongue was cut out, and he was allowed to bleed to death. Given Saddam's history of committing atrocities, this story was believable, but administration spokespersons declined to say when the incident had occurred and who had witnessed it. Amnesty International, Human Rights Watch, journalists in Baghdad, and intelligence sources were unable to confirm that it, or similar cases also cited by the administration, had actually happened.[14]

There were other missteps. When Charlotte Beers began her efforts at the State Department to "rebrand America," one of her major projects was a "shared values" television advertising campaign showcasing achievements of American Muslims. But by the time the $15 million campaign aired, the Arab world was less concerned about domestic American attitudes toward Muslims and more interested in the looming war.[15]

Beers had some successes before she resigned for health reasons in March 2003. She cited an 18-minute documentary produced by her department about the reconstruction of Afghanistan, which was shown on a news program in Pakistan. She also pointed out that the growing number of satellite television channels would be hungry for content, and the U.S. public diplomacy projects could take advantage of that by providing programs for broadcast.[16]

But it's not that simple. A skeptical and sometimes hostile public will not simply accept information that they are being spoon fed. The obviously self-serving advertising messages the United States produced seemed to some Muslims to be patronizing. Television officials even in the relatively friendly governments of Egypt and Jordan refused to air them. Only four countries—Indonesia, Kuwait, Malaysia, and Pakistan—were willing to run the "shared value" ads on their state-operated channels.[17]

America's public diplomats have seen their efforts produce only mixed results. People are working hard and lots of money is being spent, but it is clear that the overall system needs to be improved. Recommendations emerging from a number of studies call for changes, including rescuing public diplomacy programs from their scattered locations within the State Department. The theme of these suggestions is that the government's efforts need to be more cohesive and streamlined, and public diplomacy officials should be given more autonomy, with authority to eliminate duplicate ventures, improve training, expand exchange programs, and strengthen intra-governmental coordination.[18]

During the war, the government's information agenda was dominated by urgent efforts to deliver information to Iraq and the rest of the region. A joint American and British effort produced five hours of daily television programming that was broadcast into Iraq from Commando Solo, an American flying television studio in a converted Hercules transport aircraft. The British government paid an independent British media company about $17 million to produce 30 one-hour segments for "Towards Freedom TV," which would be vetted by the Foreign Office.[19] The content reflected the government's interests. When the project's news editor was asked about covering civilian casualties, he responded, "We'll do a little of that." Anti-war protests? "I don't think it's an interesting message."[20]

Proponents of this approach were not inclined to allow journalism to contradict officialdom's politically colored version of events, which meant a fundamental conflict existed between objective and partisan information providers. The Voice of America, to its credit, does present stories that U.S. officials would find unflattering, and its credibility benefits as a result. But even VOA must allow State Department review of editorials that will be interpreted as statements of official U.S. Policy. The boundary between "review" and censorship is sometimes difficult to define.

On April 10, 2003, brief addresses to the Iraqi people by George Bush and Tony Blair, subtitled in Arabic, were transmitted from Commando Solo. Bush told Iraqis that the war was bringing to an end "a brutal regime whose aggression and weapons of mass destruction make it a unique threat to the world." He promised to "respect your great religious traditions, whose principles of equality and compassion are essential to Iraq's future." Blair said, "Our forces are friends and liberators of the Iraqi people, not your conquerors, and they will not stay in Iraq any longer than is necessary. . . . Our aim is to help alleviate immediate humanitarian suffering and to move as soon as possible to an interim authority that is run by Iraqis."[21] How many Iraqis had either the electricity or the desire to watch this was not known.

Another experiment was to provide Iraqis with Arabic-dubbed versions of American network newscasts. Those Iraqis who watched were undoubtedly inspired by coverage of the Laci Peterson murder case and equally newsworthy items provided by Tom Brokaw, Peter Jennings, Dan Rather, and other members of the axis of anchors. In early May, after a month of the broadcasts, money for the project ran out. Mourning for its demise was muted.

Meanwhile within Iraq, the U.S. reconstruction office quickly began publishing an eight-page, twice-weekly newspaper, *Al-Sabah*, with a press run of 50,000 copies. Content included world news, local news, sports, and horoscopes.[22] *Al-Sabah* soon faced plenty of competition as Iraqis took advantage of their new freedom to publish their own newspapers. Before long, Iraqis had more than 100 newspapers to choose from.

While the interim U.S. efforts stumbled along, grander ventures were being planned. An early and apparently successful project was Radio Sawa. It is owned by the U.S. government and delivers news from the Voice of America, but its appeal is found in its primary content—a mix of Middle Eastern and Western music. Its target audience is the under–30 listener, and after a few months of operation in 2002, it claimed a large audience in Jordan and several of the Persian Gulf states. Sawa may eventually reach 250 million listeners.[23]

The popularity of Sawa was especially important because the Voice of America's broadcasts were only being listened to by two percent of the Arabic-speaking audience. Critics said VOA was too dry to win a broad-based audience. By mid-2002, with a $35 million budget, Sawa had set up FM stations in Jordan, Kuwait, Qatar, and the United Arab Emirates. It also had an AM facility in Kuwait from which it transmitted into Iraq. The project's plans include opening stations in a wide area, from North Africa to the Arabian Peninsula. Production quality is impressive and the news reports feature what American officials consider to be objectivity. For example, while most newscasts in the region refer to bombings by Palestinians as "martyrdom operations," Sawa calls them "suicide operations."

The audience appears to be large, but appraisals of Sawa's success should be tempered by an important concern: Among those who listen to the station's music, how many people are paying attention to, much less being influenced by, its newscasts? Some Middle Eastern critics of Sawa say not many, but neither they nor the station's proponents really know the answer.[24] Nevertheless, from the U.S. standpoint there is something to be said for having this American media presence in the region. Change in public opinion will only occur in small increments, and Sawa may help.

As for television, after the stopgap projects ran their course, U.S. officials began considering long-term ventures. President Bush included $30 million in his 2004 budget to create an Arabic-language satellite television channel, the

Middle East Television Network (METN), which would be modeled on Radio Sawa's slick and audience-friendly approach, with enough news mixed in to off-set some of the clout of Al-Jazeera and other regional broadcasters. Norman Pattiz, chairman of Westwood One, which provides content to thousands of American radio stations and other media companies, was the driving force be-hind METN, and he planned to offer a mix of original programming (includ-ing talk and children's shows) and American TV shows and movies.[25] That will probably find an audience, but, as with all such ventures, it is unclear how much impact it will have on attitudes about the United States generally and about American foreign policy in particular.

Beyond Iraq: Information Campaigns Elsewhere

The post-war scramble to shape Iraqi and global opinion has been the most vis-ible public diplomacy effort in recent years. But significant, if quieter, projects also are under way. In September 2002, the Voice of America began broadcast-ing *Next Chapter,* an MTV-style television program in Farsi directed at Iran, where 70 percent of the country's 68 million citizens are under age 30. Much of the material is subtly provocative: an interview with Jay Leno about using humor to criticize politicians; a feature about an Iranian-American college stu-dent that through her words, clothing, and lifestyle illustrates the personal di-mension of freedom. Sometimes harder-edged material is mixed in, such as videotape smuggled from Iran with people talking about the oppressiveness of the mullahs.[26]

This is how the battle goes in the ring of electronic politics: jab, jab, jab—no need to try for a knockout punch. After a while, so the theory goes, the jab-bing is likely to take effect; the opponent will be weakened by the constant barrage of information. It certainly beats conventional warfare, and the U.S. government should appreciate its value.

VOA-produced shows for Iran like *Next Chapter* and related programming on its similarly formatted Radio Farda are supplemented by private satellite television channels such as National Iranian TV, PARS TV, Azadei, and Channel One, all of which are based in Los Angeles and privately funded. Channel One provides news from Iran and takes telephone calls from Iranians who want to protest their government's policies. During the 2003 anti-government protests in Iran, Channel One anchor Shahram Homayoon stayed on the air for up to 21 hours a day, fielding calls and urging viewers to participate in the demon-strations. With the Iranian government cracking down on news media within the country, more Iranians turned to sources such as the U.S.-based channels.

The programming got the attention of Iranian leaders. Former president Ali Akbar Hashemi Rafsanjani warned Iranians to "be careful not to be trapped by the evil television networks that Americans have established," and information minister Ali Yunessi said the United States was waging psychological war against Iran.[27]

This jabbing goes on in numerous venues, government- and privately run. The Voice of America alone broadcasts in 55 languages, disseminating its material on television, radio, and the Web. Although funded by the government, VOA is guaranteed journalistic independence; its news content is not controlled by any government officials or agency. It began operations in early 1942, just after the United States entered World War II. In its first broadcast, announcer William Harlan Hale, speaking in German, told his listeners in Europe: "Here speaks a voice from America. Every day at this time we will bring you the news of the war. The news may be good. The news may be bad. We shall tell you the truth."[28]

If VOA were to develop more innovative programming such as *Next Chapter* and if that kind of implicitly political content could be disseminated more widely, the effectiveness of this form of public diplomacy would be elevated. The VOA budget for fiscal 2003 was $160 million. That money has gone a long way, and more would, of course, go still farther.

Aside from the work of VOA and other agencies, ad hoc information operations of one kind or another are almost always in progress. Some of them do not bother with subtlety. During the latter stages of the Iraq war, France complained that it was the victim of an American disinformation attack. French officials claimed that false news stories, with anonymous Bush administration officials as sources, had been used in what one French official called an "ugly campaign to destroy the image of France." American news organizations had carried stories about a variety of alleged French misdeeds: France possessed prohibited strains of smallpox virus; France had sold Iraq spare parts for aircraft and components for weapons; France had provided fleeing Iraqi leaders with French passports so they could escape to Europe.[29]

Although the French ambassador to the United States protested loudly and administration officials said that they were shocked, shocked that such accusations had been made, the controversy continued. Members of Congress called for an investigation, and Secretary Rumsfeld told reporters that "France has historically had a very close relationship with Iraq. My understanding is that it continued right up until the outbreak of the war. What took place thereafter, we'll find out."

The Bush administration wanted to harass France, and the most expeditious way to do so was through the news media. French officials were kept on the defensive while American news organizations served as de facto agents of the administration by publishing the anti-French material. Everyone involved probably understood what was going on. Journalists do not like being manipulated in this way, but there is little they can do to stop the practice other than demanding that all the charges—in this case, of French malfeasance—be on the record, with the accuser fully identified.

But unless all news organizations did this, the sources would refuse to surrender their anonymity and would seek out reporters who were not so particular. Sources can always get promises of confidentiality in exchange for juicy tidbits. What journalist can resist a story about a secret cache of smallpox virus? The losers in all this are the news consumers; they don't know whom to believe, and rarely does the press take the time to sort out the tangle of charges, countercharges, and motivations and present the public with a clear picture of what is going on.

Next Steps

The shortcomings of America's public diplomacy efforts finally received serious attention in late 2003. The Advisory Commission on Public Diplomacy, chaired by Edward Djerejian, a former U.S. ambassador to Syria and Israel, told Congress that "the United States today lacks the capabilities in public diplomacy to meet the national security threat emanating from political instability, economic deprivation, and extremism, especially in the Arab and Muslim world." The commission's report criticized "a process of unilateral disarmament in the weapons of advocacy over the last decade [that] has contributed to widespread hostility toward Americans which has left us vulnerable to lethal threats to our interests and our safety." The report called for strong coordinating leadership from the White House and for a "dramatic increase in funding" above the $600 million that had been available for public diplomacy in 2002 (of which only $25 million was for outreach programs in the Arab and Muslim world).[30]

Internal improvements in all aspects of public diplomacy are important, and so is adjusting the operating philosophy behind America's way of sending messages to the rest of the world. Distrust of U.S. intentions will not vanish quickly, and information identified as emanating from the U.S. government will probably continue to be viewed with great skepticism, if not ignored altogether. During the Iraq war, American officials worked fairly well with representatives of important Arab news organizations such as Al-Jazeera, and such efforts need to expand.

Rather than having U.S. agencies directly disseminate information and become competitors of regional news organizations in the Middle East and elsewhere, they might be better off trying to work with them while quietly encouraging them to alter the tone of their coverage of American policy and values. The United States might also provide financial support to indigenous media. This would be more cost-efficient and make more political sense because close-to-home media organizations, not distant voices, are most likely to establish credibility with the audience.

Public diplomacy can be successful only if it is part of carefully crafted foreign policy. Barry Fulton of George Washington University's Public Diplomacy Institute said: "Public diplomacy is not, and should not, be somehow considered as camouflage for public policy. Public diplomacy is describing public policy, but it doesn't improve on it, change it, or misrepresent it."[31]

In diplomacy among publics, large gaps need to be bridged. American journalists working around the world have consistently encountered skepticism, cynicism, and anger toward the United States—feelings exacerbated by the Iraq war but with deep roots that are growing deeper. At the American University of Beirut, *The Washington Post*'s David Ignatius found that the students "don't believe that America is serious about its values. Suggest to them, for example, that America really wants to advance democracy and freedom in Iraq, rather than grab the country's oil, and you get smirks and guffaws."[32]

Many young Arabs see the United States as a bastion of hypocrisy and arrogance, showing no respect toward other people and cultures. In Saudi Arabia, said Nicholas Kristof of *The New York Times*, "I kept asking women how they felt about being repressed, and they kept answering indignantly that they aren't repressed." When he asked about wearing *abayas*, one Saudi woman told him: "I cover up my body and my face, and I'm happy that I'm a religious girl obeying God's rules. I can swim and do sports and go to restaurants and wear what I want, but not in front of men. Why should I show my legs and breasts to men? Is that really freedom?" Other women made the same case to Kristof, "that Saudi women are the free ones—free from sexual harassment, free from pornography, free from seeing their bodies used to market cars and colas. It is Western women, they say, who have been manipulated into becoming the toys of men."[33]

Kristof harrumphed a bit about this, and asked why Saudi women were not at least given more of a choice about what to wear and other social issues, but in a benign way he reflected an American characteristic that aggravates much of the rest of the world: the assumption that American standards are the best and should be adopted by everyone.

Foreign policy, and public diplomacy in particular, will be weakened by incorporating that outlook. That is why criticism arose when the U.S. government printed 300,000 copies, in ten languages, of a pamphlet titled "Muslim Life in America." Muslims could rightly argue, "Now you're even trying to coopt our religion and tell us that 'the American way of Islam' is best." That was not the intent of the pamphlet—it was designed to show that Muslims are not persecuted in the United States—but regardless of intentions, ingenuousness is not a useful ingredient in public diplomacy.

In a larger policy context, if a government wants to rely on "soft power" to accomplish its goals, public diplomacy is indispensable. Joseph Nye defines soft power as "getting others to want what you want. . . . [It] is also more than persuasion or the ability to move people by argument. It is the ability to entice and attract. And attraction often leads to acquiescence or imitation." This should work in America's favor, says Nye, because "our institutions will continue to be attractive to many and the openness of our society will continue to enhance our credibility."[34]

But, warns Nye, soft power is effective only if its underlying principles are respected by policy makers as a worthwhile means of exercising influence. Nye says that "the arrogance, indifference to the opinions of others, and narrow approach to our national interests advocated by the new unilateralists are a sure way to undermine our soft power."[35]

Sometimes "hard," coercive power must be exercised, but long-term international stability is more likely to be maintained by the softer influence that is the essence of public diplomacy. Even in a time of war, ways to use news and other media to build peace are worth exploring.

CHAPTER EIGHT

The Next Wars

Crunching the numbers from the Iraq war will continue for years, but an early "analysis and assessments" report prepared in May 2003 for the Department of Defense included these items:

- U.S. military personnel deployed: 423,998
- Other coalition forces: 42,987
- War's duration: 30 days
- Coalition air sorties: 41,400
- Aircraft lost: 20; 7 due to enemy fire
- Psychological operations leaflets dropped on Iraq: 31,800,000
- Known costs up to time of report: $917,744,361.55 (equivalent of slightly less than one hour of U.S. economic output in 2001)[1]

Statistics provide a comfortably sterile view of war, but still generate their own shock and awe. The big numbers—the dollars, sorties, the propaganda leaflets—contrast strikingly with the small ones, especially the war's duration (according to the Pentagon's definition): 30 days. The figures would be even more impressive if the war had really ended.

Trying to Wrap Up Iraq

After President Bush made his *Top Gun* visit to an aircraft carrier to declare victory ("the end of major hostilities"), American news organizations emerged from their self-induced comas and began examining the administration's case for going to war. They found plenty of issues worth covering, particularly the duplicity and/or incompetence in the appraisals of Iraq's weapons of mass destruction. Bush administration officials, apparently surprised that they were being scrutinized rather than applauded, responded defensively, which gave reporters more incentive to dig.

In Iraq, continued armed resistance was scattered but persistent and lethal. It was sufficient to undermine any notion that the United States was being universally welcomed as the great liberator. The continuing deaths of American troops were proof of this. Two on one day, one the next, three the day after that; all in nasty little incidents that usually could not be called battles but were menacing and frustrating for the troops in the field, and mystifying for people in the United States who thought that the bad guys had lost and the war had ended. As if to make up for their earlier neglect in analyzing postwar prospects, news organizations intensively covered these deaths. *Time,* for instance, did a five-page spread in July 2003 focusing on one Army first sergeant who was killed. It presented shorter profiles of six others who were also killed. The *Time* report was implicitly critical of the administration for not properly preparing U.S. soldiers for the twilight war or defining their mission.[2] When criticism of strategy is wrapped in a tear-jerking story, it is more likely to capture the public's attention and cause political repercussions that policy makers must deal with.

For Americans following news coverage of events, the central part of the war—the intensive push to Baghdad—seemed relatively uncomplicated in retrospect, while the "post-war" difficulties were hard to understand. "Why do they hate us?" still had not been answered and now took on new meaning in Iraq with its bewildering array of religious and political factions. News organizations faced the task of trying to make sense of all this, and some made valiant efforts to explain the history of Iraq, the differences between Shia and Sunni Islam, and other such topics.

But the years of superficial coverage had taken their toll, and this catch-up reporting had much ground to cover just to bring the news audience up to a level of basic competence. Policy makers were affected by this because when things were perceived as going badly for U.S. forces in "postwar" Iraq, the public reacted to the latest events without considering them in the context of longer-term issues. Overall, the coverage reinforced this, devoting considerably more attention to American casualties than to the complex postwar political challenges and the occasional signs of progress in Iraq and the rest of the region.

The Bush administration had hoped that the display of American power sweeping through Iraq would have beneficial ripple effects on the behavior of Iran and Syria and on the Israeli-Palestinian conflict. Initially, those effects appeared to be slight, but President Bush made clear that he was not backing down. In a November 2003 speech, he said: "Sixty years of Western nations ex-cusing and accommodating the lack of freedom in the Middle East did nothing to make us safe, because in the long run stability cannot be purchased at the ex-pense of liberty. As long as the Middle East remains a place where freedom does not flourish, it will remain a place of stagnation, resentment, and violence ready for export."[3]

Among peripheral issues relevant to the future of the region, the status of the Kurds continued to receive little attention from the international news media. But the Kurds themselves were consolidating their political position in northern Iraq and putting new media technologies to better use, gaining trac-tion in their efforts to be noticed. Their Medya TV was up and running, and the Internet was rich in resources that could assist further development of cultural unity for the geographically dispersed Kurds. Although the Kurds have made little progress toward formal statehood, the Internet has quietly furthered the existence of a virtual Kurdistan.

The Kurds continue to be an interesting story that news organizations, es-pecially those in the United States, have difficulty addressing because (as is noted in chapter one) they fall outside the conventional boundaries of the jour-nalistic world—they are out there somewhere beyond the edge of our flat earth. During the Iraq war, there were frequent references to "Kurdish-controlled areas" and to Kurd fighters' desire to do battle against Saddam Hussein, but rarely were there adequate explanations of just who these people are, what their political interests and motivations are, and how they fit into the political and military balance of power in Iraq and in surrounding countries.

As the new politics of the region evolve, Kurdish interests will certainly be a factor, sometimes a complicating one, particularly concerning Turkey. If rela-tions between Kurds and Turks worsen, a regional concern could quickly be-come a NATO concern. It is unlikely that the American government has given much thought to preventing that or, if the situation grows more tense, dealing with it. It is even less likely that American news consumers—most of whom probably do not know that Turkey is a NATO member—would understand what a Turkey-centered crisis would be all about.

The Iraq war has left behind plenty of political debris, presenting chal-lenges to the news media. The conflict also left behind an altered press–military relationship, the most talked-about facet of which was the embedding system. Pentagon official Bryan Whitman, who helped design the embedding arrangement, said, "I think this is how we're going to cover wars in the future." Similarly, NBC executive Bill Wheatley said: "Certainly, the experience in Iraq indicates this sort of thing can work. This experiment has been successful." But

Los Angeles Times editor John Carroll was more cautious, saying he expected the Pentagon to try to tighten control over the reporters. "It's an uncontrolled flow of news," said Carroll. "That's generally okay until the war goes sour, and then it's not anything they're going to want to deal with."[4]

Between the Iraq war and the next conflict, journalists and military officials will be trying to improve their respective positions. News executives will probably seek more slots for embedded reporters and more freedom of movement for them, while Defense Department officials will probably redraft guidelines to allow more control over content and real-time reporting. The Pentagon's Victoria Clarke said that more foreign journalists will receive embedded slots the next time the system is used.[5]

John Carroll's point is important. While things are going well, the Pentagon will embrace openness, but if a conflict becomes bloodier or is less successful than anticipated, military officials can be expected to revert to traditional form and squeeze the flow of information. Several incidents occurred in the Iraq war when reporters witnessed more and described more than the Pentagon would have preferred, such as in the immediate reports about the fragging incident as the war began and in the unvarnished description of the killing of an Iraqi family at a roadblock by a U.S. unit that apparently gave the Iraqis an inadequate warning. If such stories were to dominate coverage, it is unlikely that Pentagon officials would simply shrug and say, "Well, that's a free press for you." Clamps on coverage almost certainly would be tightened.

Iraq and Fighting the Next Wars

The overall press–military relationship may become more complicated as the American approach to war continues to change. In 1991, the United States was satisfied to drive Iraqi forces out of Kuwait. In 2003, the United States conquered and occupied Iraq. In 1991, the "Powell doctrine" that mandated the use of overwhelming numbers of troops held sway. By 2003, Donald Rumsfeld, among others, was thinking differently and devising a strategy that would be overwhelming in capabilities but would emphasize speed over bulk.

Rumsfeld's original Iraq plan called for using no more than 80,000 ground troops and relying heavily on air power to pound the Iraqis into submission. General Tommy Franks, commander of the overall operation, was concerned about overestimating the effects of air strikes and proposed using 200,000 troops, which was still less than half the number deployed in 1991. The two men worked out a plan incorporating both approaches: Franks would begin the war with about 90,000 troops moving into Iraq, but with 100,000 more "in the pipeline" to be called upon as needed. As the war began, Franks said, "This will be a campaign unlike any other in history, a campaign characterized by shock, by surprise, by flexibility, by the employment of precise munitions on a scale never seen before, and by the application of overwhelming force."[6]

The "shock and awe" phase of the war was to comprise massive, precise air strikes. Refinements in air warfare continue to be made. During World War II, 3,000 sorties might be needed to knock out a single target. By the time of the Gulf war, ten sorties did the job. Today, one plane can take out ten targets.[7] In Iraq, the damage done by the U.S. air strikes surpassed planners' expectations, significantly degrading the fighting capacity of key Iraqi divisions. When the U.S. Third Infantry was able to squeeze through a gap and sprint toward Baghdad, Franks said, "All of that was made possible by air power."[8]

Another major element of the Rumsfeld-Franks plan was the extensive use of Special Operations Forces (SOF). Ten thousand SOF troops saw action in Iraq, triple the number who fought in the Gulf war.[9] During the fighting in Iraq, the SOF were integrated with other U.S. forces and appropriate tasks for each were defined. One SOF veteran of the war in Afghanistan said: "We can't take and hold ground. But there are some things we can do, and finally the civilian commanders have learned the proper mix."[10]

Among the SOF tasks early in the war were killing Iraqi sentries near the Kuwait border, finding and neutralizing Scud missile launchers, and joining CIA paramilitaries in reconnaissance missions. Special forces from coalition allies were also involved, such as the Polish commandos who seized Iraqi oil platforms. Because Turkey would not allow large-scale deployment of U.S. forces from within its borders, SOF were heavily relied on in the north and west of Iraq. Some regular U.S. units were even put under SOF command, which had rarely been done before.[11] Advances in real-time surveillance, such as from unmanned drones, have made rapid response more feasible and more important, and that is a role well-suited for SOF.[12]

Looking ahead, Rumsfeld wanted to further expand the SOF role in the overall U.S. military structure, in part by allowing SOF to propose and carry out missions on their own, rather than being appendages of regular forces units. In January 2003, Rumsfeld asked Congress to expand the SOF budget 30 percent, to about $7 billion, and to expand the number of SOF troops by about 10 percent. He also asked that SOF be "forward deployed" around the world—close to any potential action.[13]

Air power and special forces may have been the stars of the war, but the heaviest fighting was still done in traditional ways by soldiers with rifles and large numbers of armored vehicles.[14] Another factor in determining how this war was fought was the miserable performance by the Iraqi military. Iraqi troops were overmatched; they could not mount effective resistance against the technology and fighting skill of U.S. forces, and even though they were defending their homeland, Iraqi soldiers apparently had so little loyalty to Saddam Hussein that many of them surrendered or just walked away at the first opportunity. Nevertheless, U.S. military planners should resist the temptation to see the Iraq war as proof of American invincibility. Not every country's military can be counted on to collapse so promptly. It would be risky to assume, for instance, that a war against North Korea could be won by following the same strategy as was used in Iraq.

The new technology of warfare can make the transition from combat to reconstruction more difficult. Today's advanced weapons allow at least part of a war to be fought from a considerable distance, with reduced risk of incurring casualties and less direct contact with the enemy's civilian population. Then, after the major part of the war has ended, U.S. occupying forces are unprepared to exercise a softer kind of power. This was apparent in Iraq after the heavy fighting ended. The flawed transition diminished whatever constructive effect the war might have had in the region. As journalist David Ignatius noted, "An America that can actually transform the Middle East will need more Arabic speakers, social scientists who understand the Islamic world, development economists, human rights activists." He added that "real transformation will require connection, not distance."[15]

A war cannot be truly won unless postwar peace can be maintained. The go-it-alone approach of American policy makers meant that the U.S. military, virtually by itself, would have to keep order during at least the initial stages of the rebuilding of Iraq. One critic of this policy, retired general and Democratic presidential candidate Wesley Clark, argued that "the Bush administration simply failed to avail itself of the full range of tools and support that could have been made available." He added that this made the mission "far more difficult, dangerous, and open-ended than any undertaken previously."[16]

Most Americans know little about how numerous and far-flung U.S. military operations are. The research organization GlobalSecurity.org lists 86 U.S. military operations conducted from 1993 through the 2003 Iraq war. The majority of these did not involve major combat roles, but rather included American troops supporting humanitarian missions, aiding evacuation of American citizens from trouble spots, providing airlift assistance to NATO or the United Nations, and other such tasks. In early 2003, as troops were preparing for combat in Iraq, the U.S. Army Special Operations Command was deployed in 65 countries, from Colombia to Nepal.[17] Also often overlooked are the U.S. military's relationships with other countries' armies, working on counterinsurgency programs, humanitarian aid projects, and training at many levels. Sometimes these efforts generate controversy, with American officers accused of abetting corruption, human rights abuses, or other forms of malfeasance. But in many cases, American involvement of this kind produces more democratic and just behavior by foreign military than would be likely if there were no U.S. presence.[18]

The news media face the task of figuring out how to cover all of this coherently enough so the public can make sense of what is going on. Challenges arise on several levels.

- During combat operations such as those in Iraq, journalists tend to report as if they have a comprehensive view of the entire war, rather than operating on the assumption that they have just a narrow field of vision. The embedded reporters in Iraq could provide vivid descriptions of their units' actions, sometimes even live coverage of the fighting, but the excitement of covering a skirmish might distort appreciation of strategic reality. The big picture should not be neglected in favor of a series of snapshots.
- In Iraq, SOF missions and the devastatingly effective use of precision weaponry received little coverage at the time. In many instances, these facets of war do not lend themselves to on-the-scene reporting. Few journalists could safely accompany SOF troops into the intense combat environment of many of their missions, and the high-tech weapons did much of their damage in areas where reporters were not present. But there are other ways to get the story; interviews and research have value, even in the era of real-time news.
- When U.S. personnel are involved in non-combat duties such as assisting in humanitarian emergencies, news coverage tends to be skimpy, but this military role is important in shaping the world's knowledge about America and Americans' knowledge about the world. News organizations cannot be expected to cover all these situations, but more of them merit on-scene coverage, and at least the existence of such missions should be noted.
- The news media should offer closer examination of the U.S. military's behind-the-scenes work with other countries' armed forces. The American public deserves to know if the United States is advancing democracy or reinforcing anti-democratic regimes.

Beyond military operations that are exclusively American, the breadth and complexity of missions has also become an issue for NATO. Doing only what has been done in the past is a sure path to obsolescence, and NATO has come to realize that. It put its toe in the water in early 2003 by agreeing to participate in the International Security Assistance Force in Afghanistan, and in August 2003 assumed control of Afghanistan's multinational peacekeeping force. NATO secretary-general George Robertson stated that the commitment in Afghanistan proved that NATO could evolve. "After all," he said, "it was the lack of its direct involvement in the initial U.S.-led campaign there that precipitated much of the punditry about NATO's future. Now, after taking a long, hard look at itself, and launching the latest in a long line of adaptations to new circumstances," NATO would undertake its first mission outside Europe since it was formed in 1949.[19]

Robertson also noted that NATO's historic focus on conventional warfare needed revision. Given NATO's military strength, he said, "terrorism will often be our opponents' strategy of choice."[20] Similarly, Czech Republic president Vaclav Havel wrote that NATO's principal adversary of the past, the Soviet Union's "evil empire," had been replaced "by what is perhaps an even more dangerous enemy: a dispersed evil that is sophisticated and yet hard to grasp."[21]

As is the case with NATO, the United Nations finds its relationship with the United States changing significantly. If the post–9/11 approach established by George Bush remains intact, waiting for "the international community" to sanction military action will be a thing of the past, at least from an American standpoint. Bush made clear that new alliances will be forged—"coalitions of the willing" that will act rather than dither. He vowed to try "to prevent the United Nations from solemnly choosing its own irrelevance and inviting the fate of the League of Nations."[22] Longtime allies in Paris, Berlin, and elsewhere will have to decide where their interests lie, with the United States or with some new post-UN or post-NATO configuration, perhaps with the European Union at its center.

Some years will pass before all this shakes out, but the potential exists for the most dramatic revision of the Western power structure since the end of World War II. International institutions such as the UN could lose much of their relevance if the United States walks away from its commitment to making them work. American news coverage has approached these matters as largely personality-based: Bush and Blair versus Chirac and Schroder, with Putin floating on the periphery. Pronouncements such as "Punish France, ignore Germany, forgive Russia," which was attributed to national security advisor Condoleezza Rice, are catchy, reducing complex policy to soundbyte size, but the underlying issues are sophisticated and require equally sophisticated news coverage.

Globalization and War

In a speech in Chicago in 1999, Tony Blair said: "We are all internationalists now, whether we like it or not. We cannot refuse to participate in global markets if we want to prosper. We cannot ignore new political ideas in other countries if we want to innovate. We cannot turn our backs on conflicts and the violation of human rights within other countries if we want still to be secure."[23]

Globalization in its many forms is here, and, particularly for major powers, adapting to it and participating in it is compulsory. It is not a new phenomenon; interdependence has increased steadily for years, particularly during the twentieth century. This coming together has accelerated largely because of the technologies that make things happen so quickly, as information and commerce zip electronically across borders at unprecedented speed and volume.[24] *New York Times* columnist Thomas Friedman says that globalization is "the overarching international system shaping the domestic politics and foreign relations of virtually every country, and we need to understand it as such."[25]

The benefits and detriments of globalization's economic effects have received considerable scrutiny from the news media and others, but less attention has been paid to globalization's impact on conflict and conflict's effects on globalization. Kurt Campbell of the Center for Strategic and International Studies observed that "perhaps we have unintentionally subscribed to the persistent optimism of the prophets of globalization who either [implied] or explicitly stated that conditions of globalization made a major, sustained conflict most unlikely." He said that terrorism has altered any such assumptions, and asked, "Can continuing globalization and a lengthy and expensive campaign against global terrorism coexist?"[26]

Globalization will shape, and be shaped by, conflict, particularly terrorism. Conflict imposes barriers to cooperation, and fear generated by terrorism inhibits the freedom of movement—of people, goods, capital—that is part of globalization's foundation. Terrorism expert Bruce Hoffman cites changes in the United States as examples of these constraints, noting that "with every new threat to international security we become more willing to live with stringent precautions and reflexive, almost unconscious wariness." This surrendering of pieces of freedom in the interest—real or purported—of security can be debilitating. Hoffman says that terrorists seek to "shrink to nothing the areas in which people move freely," which infringes on the freedom of individuals and societal institutions. Predicting that suicide attacks will continue to be especially attractive to terrorist organizations, he says that "they guarantee media coverage. The suicide terrorist is the ultimate smart bomb. Perhaps most important, coldly efficient bombings tear at the fabric of trust that holds societies together."[27]

During the Iraq war, no American worried about a conventional attack by Iraqi armed forces against the U.S. homeland; that was inconceivable. But many

Americans worry about the cities of America becoming like the cities of Israel, where no public place seems safe. This is a difficult topic for the news media. The threat is real and it demands thoughtful examination by journalists, but news organizations must be careful not to slip into sensationalistic reporting that will frighten rather than educate the public.

This is the era of multilevel warfare. Occasionally, there will be the grand conflict of army versus army, as took place in Iraq. Much more frequent will be the terrorist sorties that are potent because of their unpredictability and the fear they produce. In between are the limited, less visible special operations wars.

Then there are the civil wars, remaining distant from much of the world but stunning in their savagery, which regardless of definitions of "national interest" can provoke a desire to "do something." Determining how to respond involves political and moral issues that the public often does not ponder until after policy makers have already decided what they will do. That is partly because the news media are slow to lay out the evidence for and against various courses of action. Australian scholar C. A. J. Coady argued that the foundation for considering such matters is the ethical question, "How should we live?" The answer to that, in terms of going to war, has been limited by respect for sovereignty. Coady said that "a just war has tended to be seen primarily as a defensive war. Military interventions in the affairs of other states without the warrant of self-defense or defense of allies were largely ruled out, both morally and legally."[28]

That has changed, partly as a function of globalization. Terrorism and humanitarian emergencies of various kinds find borders less and less restrictive. They can spread farther and faster than ever before, as can news about them. Increasingly, globalization-related factors will trump sovereignty when governments weigh reasons for intervening.

This is particularly the case for the United States, which possesses such incomparable military power that it can define the rules about going to war as it sees fit. President Bush said in November 2003 that leaders' duty to defend their people "sometimes requires the violent restraint of violent men. In some cases, the measured use of force is all that protects us from a chaotic world ruled by force."[29]

But, as Coady pointed out, having so much power can create problems. "The dominant power of the United States," he said, "is an ambiguous asset. It can help to solve problems, but, if used incautiously, it can have the effect often created by bulls in china shops. The unequaled national might of the United States is envied, feared, misunderstood, and interrogated by those who feel its impact, especially in military forms." He urged "collaborative peacekeeping and peacemaking," rather than unilateral U.S. action, even in response to human suffering that resonates with "the John Wayne lurking within us all."[30]

These are valid points concerning a decision about *whether* to intervene, but even if the decision to act is made, *how* to go about it raises more questions. Choosing to apply economic sanctions rather than taking military action was recently popular among American policy makers. The history of U.S. dealings with Iraq from 1990 to 2003 illustrates how complicated this can be.

After Iraq invaded Kuwait in 1990, the United Nations Security Council froze Iraq's foreign assets and imposed sanctions that prohibited Iraq from importing anything not expressly permitted by the UN. The Security Council also barred companies from doing business with Iraq, with a few exceptions. Before the sanctions were imposed, Iraq had been importing about 70 percent of its food, medicine, and agricultural chemicals, so the widespread belief was that the sanctions would cripple the country and force it to respond to UN demands that it leave Kuwait. That did not happen.

Then came the Gulf war, and afterward the sanctions remained in place. In 1992, the United Nations added to them, establishing "no fly" zones over some Shiite and Kurdish areas of the country. Despite the economic burden and military humiliation created by these measures, Saddam Hussein remained in power. By 1993, U.S. ambassador to the United Nations Madeleine Albright was among those who realized that the Iraqi people were suffering greatly while Hussein himself was merely inconvenienced. An oil-for-food plan was devised that allowed Iraq to sell some of its oil and use the proceeds to buy food, medicine, and a few other items. Trade in military-related goods was still barred. Hussein, however, wanted all the sanctions lifted and did not agree to the oil-for-food arrangement until 1996. Meanwhile, countries such as France and Russia wanted to do business with Iraq and were pressuring the UN to relax the sanctions.[31]

That is a quick sketch of a complex story. There was great misery for many Iraqis, much skullduggery by Hussein, and under-the-table dealing by the governments and private entities that put business above principle and evaded the trade restrictions. The bottom line was that the sanctions were intended to drive Hussein from power, and they failed to do so. Nevertheless sanctions are still embraced by some as the only alternative between doing nothing and going to war.

That sounds logical, and sanctions were considered an important tool for U.S. foreign policy makers before the attacks of September 2001. At that time, war against Iraq was not considered an option unless Hussein was foolish enough to provoke one. Since 9/11, however, war—against Iraq and, presumably, other nations—has become more accepted within the U.S. government as a policy option. That marks a profound change in America's relationship with the rest of the world, and it is recognized as such by other states. The "non, nein, nyet" trio of France, Germany, and Russia opposed the U.S. war against Iraq not simply because of the circumstances of that conflict, but also because they saw it as a dangerous precedent. The world's most powerful nation was going to use its power whenever it decided to, and other countries could either join in or get out of the way. Even Britain presumably was concerned about this, but decided it could best restrain the United States as a partner rather than an outsider.

What appears to be a profound shift in American foreign policy has attracted remarkably little attention in public forums. Congress, as is its habit on issues of war and peace, waved the flag and rolled over. News coverage focused on the crisis of the moment—Iraq—and did little to foster public debate about whether the new belligerence is appropriate as a broadly applicable course of action.

Perhaps the Iraq war will prove to be an isolated event and the threat of American military power will be enough to deter potential adversaries. On the other hand, a more menacing United States may be increasingly hated as well as feared, stimulating the kind of terrorist attacks against which massive military strength has limited effect. Speculation about this can continue indefinitely, but the public deserves to be part of a deliberative process about using armed force and deciding where America's relationship with the rest of the world is heading. News coverage is an essential tool for stimulating public interest and involvement in that process.

Those decisions must include how to deal with civil wars that are often waged outside the view of all except those directly affected by them, at least until the news media bring them into the world's field of vision. A half-century ago, civil wars were less common because many of the nations in which these conflicts occur today were then colonies and the colonial power imposed stability. But while this kind of war may have been uncommon, so was freedom.

As a growing number of former colonies became independent, they found that absent a well-designed political system, freedom in its early stages can be destabilizing. Countries often must struggle to survive the infancy and adolescence of nationhood, coping with political growing pains that in some instances prove crippling. The number of major civil wars (those with at least 1,000 battle deaths) may have declined after the Cold War, when superpowers stopped funding proxy forces, but today's civil wars drag on longer—an average of eight years.[32]

The reasons behind these internecine conflicts vary. Western press reports occasionally ascribe some of them to "tribalism," which is simplistic and may have racist connotations. The real causes tend to be political structures that give absolute power to the majority population in a country, and poverty, which when severe enough makes violence seem a reasonable path toward change. The poorest one-sixth of the world's population endures 80 percent of the world's civil wars. The World Bank reported that when income per person doubles, the risk of civil war is cut in half.[33]

When greed is added to the mix of a weak political system and pervasive poverty, the chances of civil war increase still more. "Conflict diamonds," timber, and other resources are at the root of some of these wars, especially when the rest of the world implicitly encourages the fighting by providing markets for ill-gotten prizes. Recognizing this, 45 nations engaged in the diamond trade agreed in 2002 to not accept conflict diamonds for cutting and selling. Depriving warriors of international outlets for their loot is helpful, but stimulating legitimate economic growth may be the best way to dampen the fires of civil war, which the World Bank has called "development in reverse."[34]

A number of African countries are among those with the conditions in which civil war festers. Between 1990 and 2001, per capita income in sub-Saharan Africa declined by 0.2 percent. The number of people living in poverty continues to increase, and life expectancy continues to decline. This latter characteristic is accelerating because of the AIDS epidemic. In much of Africa, the economy is moribund, partly because Africa attracts only about 1 percent of the world's foreign direct investment, even though 10 percent of the world's people live there.[35]

The economy also suffers because of policies of first-world nations such as the United States that effectively block access to world markets for African products such as cotton. Protests against these government agricultural subsidies brought an end to the World Trade Organization meeting in Cancun, Mexico, in 2003. That kind of militancy will certainly continue and expand, increasing tensions in the already volatile relationships between powerful and developing nations.

Reducing conflict in Africa will require, among other measures, making globalization a beneficial reality throughout the continent. In 2003, the Commission on Capital Flows to Africa stressed the importance of a steady stream of private capital. It recommended a tax break for U.S. companies investing in Africa: no taxes on repatriated profits for ten years. This incentive could boost investment by more than a billion dollars annually. The commission also recommended that the United States negotiate free trade agreements with individual African nations and that all African products be allowed to enter the United States duty-free and without quotas. This latter measure, said David Ignatius, would be "discriminatory and unfair to other nations, but so what? Africa needs some positive discrimination."[36]

Telling more of the world about events in Africa and other often-ignored parts of the world could also be helpful. Governments, NGOs, and even individuals might be more disposed to address Africa's problems if they were better informed about them.

Globalization is not really global if it leaves entire continents behind. Critics of globalization have argued that the term itself is deceptive, masking just another way for the rich to get richer. Such an imbalance could easily happen, but the tools for true global economic growth do exist. The chances of their being put to work might be better if the public understood what is involved. The basic economic principles are not all that complicated, and some good explanatory journalism might awaken people to the importance and feasibility of global economic justice.

But meanwhile, the fighting goes on. Wars in Central and West Africa have been particularly persistent and horrific. Congo is one sad example that illustrates how a civil war can quickly become a regional conflict. In 1998, as Congo went through paroxysms of domestic unrest, Rwanda and Uganda invaded to dislodge the Congolese government. Zimbabwe, Angola, and Namibia then sent troops to prop up the regime. All the invaders helped themselves to Congo's resources, local militias rampaged, and then erstwhile allies Rwanda and Uganda started fighting each other. Out of a population of 55 million, well over three million had died by mid-2003 as a result of the war, the vast majority from disease or starvation.[37]

With fighters and refugees pouring over borders, and the economies and politics of the region destabilized, the United Nations estimated that Congo's war has affected 16 million people. Watching Congo implode, Philip Gourevitch, chronicler of much African misery, wrote: "Oh, Congo! What a wreck. It hurts to look and listen, and it hurts to turn away."[38]

Among the combatants in the wars of sub-Saharan Africa are roughly a quarter-million *kadogos*—"small ones" in Swahili, the child soldiers, some as young as seven, who are used by their elders as fodder to satisfy the appetite of war. They are given guns and beer and told to kill. The prospect of facing one of these small but deadly soldiers haunts peacekeepers, according to a UNICEF official who added, "It's horrible, but it shows to what level of craziness and inhumanity this has come to." International law makes it a war crime for a military force to use soldiers younger than 15, but international law has little clout in these war zones.[39]

The use of child soldiers is sensationally horrible, and therefore gets some attention from the news media. Newspapers and television have carried plenty of pictures of sweet-looking children wielding AK–47s. The meaning of the images sinks in only if those who see them suspend their disbelief and recognize the level of moral corruption represented by this exploitation.

Then the questions arise. What could be done to stop this? Can and should the United States play a role?

For more than a decade, most American involvement and, more often, noninvolvement in response to African crises has ranged from farcical to pathetic, with consistency only in the lack of effectiveness. The retreat from Somalia in 1993, the failure to act in Rwanda in 1994, the indecisiveness about acting in Liberia in 2003, the hesitation in committing resources to battle AIDS throughout the continent—these and other examples constitute a history of malign neglect. There was coverage of these events in American and other first-world media, but it often was belated, superficial, or limited to elite news organizations that reach a knowledgeable but relatively small audience.

On the military side, a few constructive steps have been taken. In 1996, the Clinton administration allocated $35 million to organize an African peacekeeping force of slightly less than 10,000 troops from African countries. This would not be a standing army, but the components could be deployed quickly from their bases in their home countries. After several name changes, the program was designated African Contingency Operations Training and Assistance by the Bush administration in 2002, and its mission was expanded to include training for offensive military operations—light infantry and small unit—as well as traditional peacekeeping skills. The basic purpose is to let Africans take care of Africa. That can be viewed as a useful assist to regional autonomy, or as a way for America to avoid getting directly involved in Africa.

Another initiative was the 2003 plan of the 15 governments of the European Union to send a peacekeeping force to Congo. The 1,400-member force was authorized by the United Nations and was the first EU military operation that did not involve NATO, as well as the first EU military operation outside Europe. France initiated the mission, presumably to show European resolve in the absence of an American response.[40]

Yet another option that has been considered is to hire soldiers to do military chores in Africa. Employees of "private military companies" do not particularly like being called "mercenaries," but whatever they are called, their business is booming. There are so many low-intensity conflicts in progress around the world at any given time that there is plenty of work for firms such as Military Professional Resources, Dyncorp, Combat Support Associates, and Levdan. They are employed by governments and corporations, and may be called upon to protect relief organizations and other NGOs.[41] Kofi Annan has said that even the United Nations has contemplated using them as a rapid reaction force for peacekeeping missions. But, said Annan, "the world may not be ready to privatize peace."[42]

Of course, if terrorists have privatized war, not much of a leap is required to see the logic in privatizing some aspects of peacekeeping. War in its smaller manifestations could conceivably become the preserve of non-state actors.

Much is going on, and where are the news media? Journalists flock to the major wars that feature shock and awe, but what about the wars in which the machete rather than the Cruise missile is the weapon of choice? Several thousand Iraqis and a few hundred Americans died in the Iraq war, while millions have died in Congo, Sierra Leone, Sudan, and elsewhere in Africa, which is just one locus of these "little" wars.

Ideally, given the expanded venues for news, there could be a growing "CNN awareness" about human rights issues, as news organizations use various media to cover conflict. This, said NATO's George Robertson, could kindle a "potent mix of anger and compassion."[43] For reaching a large and broad-based audience quickly, television still has special power, which is sometimes employed usefully. Michael Ignatieff said that "television has become the privileged medium through which moral relations between strangers are mediated in the modern world." He added, "Television's good conscience could be described [as] to pay attention to the victims, rather than the pieties of political rhetoric; to refuse to make a distinction between good corpses and bad ones . . . ; and to be a witness, a bearer of bad tidings to the watching conscience of the world." But, Ignatieff cautioned, although television might contribute to "the internationalization of conscience," its "gaze is brief, intense, and promiscuous."[44]

The fickleness of television and the news media collectively is partly a function of their lack of true independence. If government decides that a particular war or other crisis does not matter much, the news media will rarely quarrel with that. They will look elsewhere, following government's lead, for stories to tell. Policy makers are averse to bad news, and so are media organizations that worry about their audience's reluctance to face the world's grim realities. Media organizations and the public are complicit in refusing to acknowledge the existence of places such as the parts of Africa where bad news is often the only news. So instead of the news media presenting reality, the public gets "reality" television shows on which collections of exhibitionists behave outrageously for a sizable audience of voyeurs.

There are exceptions. The roll call of elite news products is distinguished although short. *The Washington Post, The New York Times, The Economist,* ABC's *Nightline,* PBS's *The NewsHour,* NPR's news programs, the occasional documentary, and a growing number of Web sites are among those that deserve to be on the list. But when compared to the overall amount of programming generated by the various media, the output of the best is too small to not be overwhelmed by the flood of mediocrity.

Covering the World of Post-Modern War

Anyone thinking that the 2003 Iraq war might mark a lasting turnaround in international news coverage will probably be disappointed. Coverage of Iraq around the time of the 1991 Gulf war is a reasonable precedent, and the Tyndall Report found that network news coverage of Iraq went from 1,177 minutes during January 1991 to 48 minutes in August of that year.[45]

Coverage of Afghanistan is another example of the short attention span of many news organizations. According to the Tyndall Report, in November 2001, Afghanistan received 306 minutes of coverage; in January 2002, 106 minutes; in February 2002, 28 minutes; in January 2003, 11 minutes; in March 2003, one minute. Comparable declines occurred in American newspapers, and the dropoff is more precipitous if the coverage appearing in *The New York Times* and *The Washington Post* is excluded.[46]

Neglect of the Afghanistan story illustrates the U.S. news media's flawed approach to international events. The coverage reflects little understanding that the conduct of international relations—including war—is an incremental process. A war does not begin and end with ferocious, ready-for-prime-time combat. It has a prelude and an aftermath that are important, even if less dramatic. No conflict erupts spontaneously; there are reasons for it that are manifest well in advance of the fighting. And afterward, peace must be sustained, and the human and material costs of the war must be addressed through rebuilding on many levels. Covering these matters is far more complicated than putting a camera on top of a building and watching Baghdad be blown apart.

By mid-2003, more than a year after the Taliban had been driven out of power in Afghanistan, Osama bin Laden's whereabouts remained unknown, more than 9,000 American troops remained in the country, and it was far from clear whether Afghanistan was going to emerge as a functioning state or slip back into being an anarchic incubator of terrorism. That is an important story, but only a few news organizations bothered to seriously cover it. John Schidlovsky of the Pew International Journalism Program said, "The public has a perception that the war is over, and then the media begin to take correspondents away and that fuels the public's perception that the war is over, and so on."[47]

One of the few American news organizations to maintain interest in Afghanistan was the *St. Petersburg Times,* which sent reporter Chuck Murphy there during the Iraq war. Murphy noted that the small number of correspondents in the country meant that the public would be able to see events from just a few viewpoints. "It's a dilution of voices," said Murphy, "and I think ultimately the readers aren't being served as well."[48]

Although the situation in Afghanistan may be complicated, the decision whether to cover events there is relatively straightforward. News organizations must determine where they want to put their resources: should they cover Afghan president Hamid Karzai as he tries to salvage his country, or focus on basketball star Kobe Bryant as he tries to evade a rape charge? Part of the decision-making about such matters is grounded in news executives' lack of respect for their audience's intelligence. They assume that the public is more interested in Kobe than Karzai, which is undoubtedly true because Kobe gets covered and Karzai doesn't. News organizations regularly abandon their responsibility to make independent, informed judgments about what the public *needs* to know.

In the case of Iraq, abdication of professional responsibility after the heaviest fighting is also related to the cheerleading and passivity that characterized much of the coverage before and during the war. In the summer of 2003, when the press finally challenged the quality of information that the Bush administration had used to justify the war, some journalists recognized that they had not done their jobs earlier. *Washington Post* ombudsman Michael Getler was among those asking tough questions: Had the news media taken adequate note of the administration's shift in emphasis from events in Afghanistan to those in Iraq? Had the press recognized the change in the level of domestic dissent when the battlefield shifted to Iraq? Had news organizations thoroughly analyzed the government's case concerning Iraqi weapons of mass destruction?[49]

Getler noted that pursuing these stories requires overcoming problems such as the shortage of sources willing to speak on the record, making it necessary to rely on those whose identity is not revealed to the public. He said that "although reporters may trust these sources and may have checked their information with other sources, quoting people anonymously still erodes the confidence of some readers and gives administration spokesmen an edge in the battle for making a case and molding attitudes." Despite that, said Getler, those sources should be used when necessary. "News organizations," he contended, "must go all-out to get answers, to guard against being used, to test their information with multiple sources and to get as much on the record as they can. And readers must be more willing not to automatically reject stories that use unnamed sources. This is a tough story to report, and information needs to get out any way it can."[50]

Getler was correct to note the *readers'* responsibilities. Members of the public should not expect that the news will be presented to them in nice, neat packages, with all loose ends tied up. Rarely do news stories emerge that way from the reporting process. Anyone who has been a news consumer for a while understands how the game is played: the occasional (but not constant) use of unnamed sources; the ambiguous wording from "official spokespersons"; the prevalence of one point of view on one day and the other side on another; and so forth. Delivering the news is an imperfect craft, and the public must consider that when making judgments about news coverage and about the events, people, and policies being covered.

───

The process of gathering and presenting news is not a purely mechanical function, regardless of the high-tech tools available. It is dependent on people—the women and men who do journalism. The vast majority of them work hard and try to deliver coverage that is meaningful and fair.

Their work is rarely easy, and, when they are covering a war, it is often dangerous. Post-modern war and the coverage of it might be best known for the glitzy technology employed by the military and news organizations, but as in wars past the real story is that lots of people die: soldiers, civilians living in the war zone, and—somewhere between those two categories—journalists.

Sixteen of them died in Iraq during the first month of that war, and more will die getting the news in wars still to come.

───

Wars still to come. Those words hold an ominous but inescapable truth. The Iraq war terminated Saddam Hussein's regime, but was the end of nothing else. Conflict may have entered its post-modern era, but the evil at war's heart is little changed.

As they move ahead in their respective roles, the policy makers, the military, the public, and the news media should recognize this, but not accept it. They must try to do better at meeting their respective responsibilities, and retain hope that war will someday cease to dominate so many aspects of politics, news, and the rest of human life.

NOTES

A Note About Online Sources

Many sources cited in the end notes and listed in the bibliography were found on line, on news and other Web sites. Although some of the Web addresses remain valid, since the research for this book was done, some of the specific URLs have changed. Some of the material may now be available only for a fee from archives, and some of it may have been pulled off the Web site altogether. Different site proprietors follow different procedures. Nevertheless, the citations should prove helpful for finding the source material if the reader is willing to engage in a bit of electronic detective work.

Chapter One

1. Howard Kurtz, "For Media After Iraq, A Case of Shell Shock," *Washington Post*, April 28, 2003, A1.
2. "Foreign News Fluctuations," *American Journalism Review*, vol. 25, no. 7, October/November 2003, 37.
3. Lucinda Fleeson, "Bureau of Missing Bureaus," *American Journalism Review*, vol. 25, no. 7, October/November 2003, 34.
4. Dwight L. Morris & Associates, "America and the World: The Impact of September 11 on U.S. Coverage of International News," survey conducted for the Pew International Journalism Program, June 2002, 3.
5. "America and the World," 9.
6. "America and the World," 12.
7. "America and the World," 3.
8. "America and the World," 17.
9. "America and the World," 13, 14.
10. Project for Excellence in Journalism, "The War on Terrorism: The Not So New Television News Landscape," May 23, 2002, www.journalism.org/resources/reports/landscape, 2.
11. "America and the World," 22.
12. Peter Singer, *One World* (New Haven: Yale University Press, 2002), 8.
13. Kenichi Ohmae, "The Rise of the Region State," in *Globalization and the Challenges of a New Century*, edited by Patrick O'Meara, Howard D. Mehlinger, and Matthew Krain (Bloomington, IN: Indiana University Press, 2000), 93, 95.
14. Samuel P. Huntington, *The Clash of Civilizations and the Remaking of World Order* (New York: Simon and Schuster, 1996), 21.
15. Huntington, *The Clash of Civilizations*, 34.
16. Charles A. Kupchan, *The End of the American Order* (New York: Knopf, 2002), 68.

17. Bernard Lewis, "'I'm Right, You're Wrong, Go to Hell,'" *Atlantic Monthly*, May 2003, 36.

18. Benedict Anderson, *Imagined Communities* (London: Verso, 1991), 184.

19. Robert D. Kaplan, *The Coming Anarchy* (New York: Random House, 2000), 38–9.

20. Kaplan, *The Coming Anarchy*, 40.

21. Kaplan, *The Coming Anarchy*, 44.

22. Robert D. Kaplan, *The Ends of the Earth* (New York: Vintage, 1997), 337.

23. John L. Esposito, *Unholy War: Terror in the Name of Islam* (New York: Oxford, 2002), 75, 79.

24. Esposito, *Unholy War*, 39.

25. Gary R. Bunt, *Islam in the Digital Age* (London: Pluto Press, 2003), 211.

26. Margaret MacMillan, *Paris 1919* (New York: Random House, 2002), 132.

27. Niall Ferguson, *Empire* (New York: Basic Books, 2003), 223.

28. Philip Bobbitt, *The Shield of Achilles: War, Peace, and the Course of History* (New York: Knopf, 2002), xvi.

29. Jon Lee Anderson, *The Lion's Grave: Dispatches from Afghanistan* (New York: Grove, 2002), 68–9.

30. Av Westin, *Newswatch* (New York: Simon and Schuster, 1982), 22.

31. Richard Reeves, *President Kennedy: Profile of Power* (New York: Simon and Schuster, 1993), 209.

32. Reeves, *President Kennedy*, 212.

33. Philip Seib, *The Global Journalist: News and Conscience in a World of Conflict* (Lanham, MD: Rowman and Littlefield, 2002), 27–9.

34. Susan D. Moeller, *Compassion Fatigue* (New York: Routledge, 1999), 153.

35. James A. Baker III, *The Politics of Diplomacy* (New York: Putnam's, 1995), 520.

36. Mohammed el-Nawawy and Adel Iskandar, *Al-Jazeera* (Boulder, CO: Westview, 2002), 34.

37. Daniel J. Wakin, "Online in Cairo, With News, Views, and 'Fatwa Corner,'" *New York Times*, October 29, 2002, 3.

38. Todd S. Purdum, "Russian and Chinese Subscribers Rely on Digest of U.S. News," *New York Times*, February 2, 2003, A4.

Chapter Two

1. Mary Kaldor, *New and Old Wars: Organized Violence in a Global Era* (Cambridge, UK: Polity Press, 1999), 8.

2. Stephen Budiansky, "The Weakness of Air Power," *Washington Post National Weekly Edition*, December 24, 2001, 26.

3. Eric Schmitt, "Improved U.S. Accuracy Claimed in Afghan Air War," *New York Times*, April 9, 2002, A14.

4. John Keegan, *War and Our World* (New York: Vintage, 2001), 58.

5. Kaldor, *New and Old Wars*, 110.

6. Robert D. Kaplan, "A Tale of Two Colonies," *Atlantic Monthly*, April 2003, 46.

7. John Lewis Gaddis, "Setting Right a Dangerous World," *Chronicle of Higher Education*, January 11, 2002, B10.

8. Steven Simon, "The New Terrorism," *Brookings Review*, Winter 2003, 18.

9. Bruce Hoffman, *Inside Terrorism* (New York: Columbia University Press, 1998), 209.

10. Simon, "The New Terrorism," 23.

11. Michael Ignatieff, "The Burden," *New York Times Magazine,* January 5, 2003, 26.

12. "Walter Rockler," *Economist,* March 23, 2002, 82.

13. Michael Glennon, *Limits of Law, Prerogatives of Power: Interventionism After Kosovo* (New York: Palgrave Macmillan, 2001), 209.

14. Robert G. Kaiser and Keith B. Richburg, "NATO Looking Ahead to a Mission Makeover," *Washington Post,* November 5, 2002, A1.

15. Ignatieff, "The Burden," 25.

16. Robert Kagan, *Of Paradise and Power* (New York: Knopf, 2003), 97.

17. Charles Kupchan, *The End of the American Era* (New York: Knopf, 2002), 117.

18. Michael Ignatieff, "Why Are We in Iraq? (And Liberia? And Afghanistan?)" *New York Times Magazine,* September 7, 2003, 72.

19. George W. Bush, "New Threats Require New Thinking," in Micah L. Sifry and Christopher Cerf, *The Iraq War Reader* (New York: Touchstone, 2003), 269.

20. "The National Security Strategy of the United States of America," http://www.whitehouse.gov/nsc/print/nssall, accessed September 30, 2002, 1.

21. John Lewis Gaddis, "A Grand Strategy of Transformation," *Foreign Policy,* November/December 2002, 56.

22. Nicholas Lemann, "The Next World Order," *New Yorker,* April 1, 2002, 45–46.

23. Ignatieff, "The Burden," 28.

24. Kenneth M. Pollack, *The Threatening Storm: The Case for Invading Iraq.* (New York: Random House, 2002), 411–424.

25. Joseph S. Nye, "Before War," *Washington Post,* March 14, 2003, A27.

26. Richard Falk, "The New Bush Doctrine," in Sifry and Cerf, *The Iraq War Reader,* 273, 277.

27. "ABC News Poll: War and the Media," ABCNews.com, January 17, 2003.

28. Felicity Barringer, "'Reality TV' About G.I.s on War Duty," *New York Times,* February 21, 2002, C1.

29. Chris Hedges, *War Is a Force That Gives Us Meaning* (New York: PublicAffairs, 2002), 143.

30. Brian Lambert, "TV Coverage of War on Terrorism Heavy on Opinion, Short on Facts," *Houston Chronicle,* February 5, 2002, 6.

31. Mark Jurkowitz, "Fighting Terror/The Home Front: The Media," *Boston Globe,* January 28, 2002, A9.

32. Susan Sontag, "First Reactions," *The New Yorker,* September 24, 2001, 42.

33. Bethami A. Dobkin, *Tales of Terror: Television News and the Construction of the Terrorist Threat* (Westport, CT: Praeger, 1992), 52, 111.

34. Brigitte Nacos, *Mass-mediated Terrorism* (Lanham, MD: Rowman and Littlefield, 2002), 16.

35. Tim Cuprisin, "Bias Seen No Matter What Term Media Use," *Milwaukee Journal Sentinel,* October 30, 2002, 12B.

36. Vasily Bubnov, "Are Chechens 'Rebels' or 'Terrorists'?" *Pravda,* October 29 2002, http://english.pravda.ru/main/2002/10/29/38856 (accessed 10/30/02).

37. Dobkin, *Tales of Terror,* 108.

38. Alina Tugend, "Explaining the Rage," *American Journalism Review,* vol. 23, no. 10, December 2001, 26.

39. Guy Keleny, "Mea Culpa: One Person's Terrorist Is Another's Freedom Fighter," *The Independent,* April 20, 2002, 17.

40. Philip Seib, ed., *New Wars, New Media: Covering Armed Conflict Since the Gulf War* (Milwaukee: Marquette University, 2001), 16.

41. Walter Laqueur, *No End to War* (New York: Continuum, 2003), 236.
42. Michael Getler, "The Language of Terrorism," *Washington Post,* September 21, 2003, B6.
43. Eric Lawee, "Palestinian Terrorists? Not on CBC," *National Post,* July 19, 2002, A 16.
44. Nacos, *Mass-mediated Terrorism,* 152.
45. Jason Burke, *Al-Qaeda* (London: I. B. Tauris, 2003), 18.
46. Michael Getler, "Here's Some News, But Be Careful," *Washington Post,* December 22, 2002, B6.
47. George Kennedy, "The British See Things Differently," *Columbia Journalism Review,* March/April 2002, 49.
48. David Greenberg, "We Don't Even Agree on What's Newsworthy," *Washington Post,* March 16, 2003, B1.
49. Virginia Woolf, *Three Guineas* (New York: Harcourt, 1938), 6, 11.
50. Susan Sontag, *Regarding the Pain of Others* (New York: Farrar, Straus and Giroux, 2003), 12–13.
51. Sontag, *Regarding the Pain of Others,* 14.
52. Philip Kennicott, "The Illustrated Horror of Conflict," *Washington Post,* March 25, 2003, C1.
53. Kennicott, "The Illustrated Horror," C1.
54. Chris Hedges, "The Press and the Myths of War," *The Nation,* April 21, 2003.
55. W. G. Sebald, *On the Natural History of Destruction* (New York: Random House, 2003), ix.

Chapter Three

1. Peter Braestrup, *Big Story* (Novato, CA: Presidio, 1994), x.
2. Colin Powell, *My American Journey* (New York: Random House, 1995), 123.
3. Lyndon Baines Johnson, *The Vantage Point* (New York: Holt, Rinehart, and Winston, 1971), 384.
4. Clark Clifford, *Counsel to the President* (New York: Random House, 1991), 474.
5. Julie Salamon, "New Tools Make War Images Instant but Coverage No Simpler," *New York Times,* April 6, 2003, B13.
6. Tom Goldstein, *The News at Any Cost* (New York: Simon and Schuster, 1985), 32.
7. Robert Harris, *Gotcha!: The Media, The Government and the Falklands Crisis* (London: Faber and Faber, 1983), 62.
8. Glasgow University Media Group, *War and Peace News* (Milton Keynes, UK: Open University Press, 1985), 9.
9. Michael Cockerell, *Live from Number 10* (London: Faber and Faber, 1988), 275.
10. David R. Gergen, "Diplomacy in a Television Age: The Dangers of Teledemocracy," in Simon Serfaty, ed., *The Media and Foreign Policy* (New York: St. Martin's Press, 1991), 59.
11. Frank Aukofer and William P. Lawrence, *America's Team: The Odd Couple* (Nashville: Freedom Forum First Amendment Center, 1995), 194.
12. Everette E. Dennis, David Stebenne, John Pavlik, Mark Thalhimer, Craig LaMay, Dirk Smillie, Martha FitzSimon, Shirley Gazsi, and Seth Rachlin, *The Media at War* (New York: Gannett Foundation Media Center, 1991), 1.
13. Jason DeParle, "Long Series of Military Decisions Led to Gulf War News Censorship," *New York Times,* May 5, 1991, A1.

14. David Lamb, "Pentagon Hardball," *Washington Journalism Review*, vol. 13, no. 3, April 1991, 33.

15. Walter Cronkite, "What Is There To Hide?" *Newsweek*, February 25, 1991, 43.

16. Jon Van, "Satellites Signal New Era in News Coverage, Viewing," *Chicago Tribune*, March 22, 2003, 1.

17. S. Abdallah Schliefer, "Interview with Ian Ritchie," *Transnational Broadcasting Studies Journal*, No. 10, Spring-Summer 2003, www.tbsjournal.com/ritchie.

18. Michael Murrie, "New Technology Brings Live Coverage of the War in Iraq," *Communicator*, May 2003, 8.

19. Paul Friedman, "TV: A Missed Opportunity," *Columbia Journalism Review*, May/June 2003, 30.

20. Steven Livingston, "The New Media and Transparency: What Are the Consequences for Diplomacy?" in Evan H. Potter, ed., *Cyber-Diplomacy* (Montreal: McGill-Queen's University Press, 2002), 118.

21. Bob Sullivan, "Expect To Watch This War From Above," MSNBC, February 21, 2003, www.msnbc.com/news/872612.

22. Maryann Lawlor, "We All Live in a See-All World," *Signal Magazine*, January 2003, www.us.net/signal/SourceBook/we-jan.

23. Cheryl Johnston, "Digital Deception," *American Journalism Review*, May 2003, 10.

24. "Public Affairs Guidance on Embedding Media During Possible Future Operations/Deployments in the U.S. Central Command's Area of Responsibility," United States Department of Defense, www.defenselink.mil/news/Feb2003/d20030228pag, 1–2.

25. Katherine M. Skiba, "Journalists Embodied Realities of Iraq War," *Milwaukee Journal Sentinel*, September 14, 2003, J1.

26. John Laurence, "Embedding: A Military View," *Columbia Journalism Review*, March/April 2003, www.cjr.org/year/03/2/webspecial.

27. Salamon, "New Tools Make War Images Instant," B13.

28. Peter Robertson, "Free Ride," *Milwaukee Magazine*, July 2003, 22.

29. Howard Kurtz, "The Ups and Downs of Unembedded Reporters," *Washington Post*, April 3, 2003, C1.

30. Michael Massing, "The High Price of an Unforgiving War," *Columbia Journalism Review*, May/June 2003, 35.

31. Laurence, "Embedding: A Military View."

32. Jacqueline E. Sharkey, "The Television War," *American Journalism Review*, vol. 25, no. 4, May 2003, 20.

33. Mercedes M. Cardona, "TV Ad Spending Rebounds During Week Three of War," *AdAge.com*, April 23, 2003, www.adage.com/news.cms?newsId=37683.

34. Lawrence K. Grossman, "War and the Balance Sheet," *Columbia Journalism Review*, May/June 2003, 6.

35. Project for Excellence in Journalism, "Embedded Reporters: What Are Americans Getting?" April 3, 2003, www.journalism.org/resources/briefing/archive/war/embedding/default, 1, 12.

36. "Embedded in Iraq: Was It Worth It?" *Washington Post*, May 4, 2003, B3.

37. Gordon Dillow, "Grunts and Pogues: The Embedded Life," *Columbia Journalism Review*, May/June 2003, 33.

38. Scott Bernard Nelson, "Embedded Reporter Comes Away from Front Lines Torn," *Boston Globe*, April 22, 2003, E1.

39. Dillow, "Grunts and Pogues," 33.

40. Peter Baker, "Inside View," *American Journalism Review,* May 2003, 39.

41. Sherry Ricchiardi, "Close to the Action," *American Journalism Review,* May 2003, 32.

42. Kurtz, "The Ups and Downs of Unembedded Reporters," C1.

43. Baker, "Inside View," 39.

44. Justin Lewis, "Facts in the Line of Fire," *Guardian,* November 6, 2003.

45. Cheryl Johnston, "Casualties of War," *American Journalism Review,* May 2003, 35.

46. Kurtz, "The Ups and Downs of Unembedded Reporters," C1.

47. Skiba, "Journalists Embodied Realities of Iraq War," J3.

48. Bill Carter, "Nightly News Feels Pinch of 24-Hour News," *New York Times,* April 14, 2003, C1.

49. Sharkey, "The Television War," 20.

50. Jim Rutenberg and Bill Carter, "Spectacular Success or Incomplete Picture?" *New York Times,* April 20, 2003, A1.

51. Salamon, "New Tools Make War Images Instant," B13.

52. Project for Excellence in Journalism, "Embedded Reporters: What Are Americans Getting?," 9.

53. Rutenberg and Carter, "Spectacular Success or Incomplete Picture?" A1.

54. Brian Lowry and Elizabeth Jensen, "The 'Gee Whiz' War," *Los Angeles Times,* March 28, 2003, A1.

55. Greg Mitchell, "Fifteen Stories They've Already Bungled," *Editor & Publisher Online,* March 27, 2003, www.editorandpublisher.com/editorandpublisher/headlines/article_display.jsp?vnu_content_id=1850208.

56. Project for Excellence in Journalism, "Embedded Journalists: What Are Americans Getting?" 10–11.

57. Rutenberg and Carter, "Spectacular Success or Incomplete Picture?" A1.

58. Alex Neill, "Successful Media Experiment Led to 'Interesting Dynamic,' Brooks Says," *Army Times,* April 22, 2003, www.armytimes.com/print.php?f=1–292925–1793248.php.

59. Thomas Kunkel, "Rushing to Judgment," *American Journalism Review,* vol. 25, no. 4, May 2003, 4.

60. Project for Excellence in Journalism, "Embedded Reporters: What Are Americans Getting?" 4–5.

61. Steve Johnson, "Realities of War Put News Ethics to Test," *Chicago Tribune,* March 26, 2003.

62. Howard Kurtz, "Photo Illustrates Rift Between Army, *Army Times,*" *Washington Post,* May 5, 2003, C1.

63. Sharkey, "The Television War," 21.

64. Joe Strupp, "How Papers Are Covering Iraqi Civilian Casualties," *Editor and Publisher,* April 8, 2003, www.editorandpublisher.com/editorandpublisher/headlines/article_display.jsp?vnu_content_id=1859190.

65. G. Jefferson Price III, "When Editors Opt Out of War's Truth," *Baltimore Sun,* April 16, 2003, A16.

66. Robert Collier, "Why I'm Still Alive," *Dangerous Assignments,* Spring/Summer 2003, 10.

67. Anne Garrels, *Naked in Baghdad* (New York: Farrar, Straus and Giroux, 2003), 125, 204.

68. "Cox's Craig Nelson on the Truth from Baghdad," *Editor and Publisher,* April 8, 2003, www.editorandpublisher.com/editorandpublisher/headlines/article_display.jsp?vnu_content_id=1859189.

69. Pew Research Center for the People and the Press, "War Coverage Praised, But Public Hungry for Other News," April 9, 2003, people-press.org/reports/print.php3?PageID=697.

Chapter Four

1. Jonathan Weisman, "Open Access for Media Troubles Pentagon," *Washington Post*, March 31, 2003, A25.
2. Weisman, "Open Access for Media Troubles Pentagon," A25.
3. Thomas E. Ricks, "Rumsfeld, Myers Again Criticize War Coverage," *Washington Post*, April 18, 2003, A28.
4. "Assessing Media Coverage of the War in Iraq," Brookings Institution forum, June 17, 2003, www.brookings.edu/comm/events/20030617.pdf, 33.
5. "By the Numbers," *Columbia Journalism Review*, May/June 2003, 27.
6. David Zurawik, "Barrage of Military Opinions," *Baltimore Sun*, March 31, 2003, 1C.
7. Ricks, "Rumsfeld, Myers Again Criticize War Coverage," A28.
8. Jim Rutenberg, "Ex-Generals Defend Their Blunt Comments," *New York Times*, April 2, 2003, B9.
9. "Assessing Media Coverage," 17.
10. Rutenberg, "Ex-Generals Defend Their Blunt Comments," B9.
11. Rutenberg, "Ex-Generals Defend Their Blunt Comments," B9.
12. Peter Johnson, "Military Experts Draw Unfriendly Fire," *USA Today*, April 3, 2003, www.usatoday.com/life/world/iraq/2003–04–03-military-experts_x.
13. Jill Rosen, "The TV Battalion," *American Journalism Review*, vol. 25, no. 4, May 2003, 50.
14. Pew Research Center, "War Coverage Praised, But Public Hungry for Other News," 1.
15. Terence Smith, "Hard Lessons," *Columbia Journalism Review*, May/June 2003, 28.
16. "Assessing Media Coverage," 27.
17. John MacArthur, "The Lies We Bought," *Columbia Journalism Review*, May/June 2003, 62.
18. Amy Tubke-Davidson, "War and Intelligence," www.newyorker.com/online/content/?030512on_onlineonly01.
19. Susan Schmidt and Vernon Loeb, "'She Was Fighting to the Death': Details Emerging of W. Va. Soldier's Capture and Rescue," *Washington Post*, April 3, 2003, A1.
20. Michael Getler, "Reporting Private Lynch," *Washington Post*, April 20, 2003, B6.
21. Michael Getler, "A Long, Incomplete Correction," *Washington Post*, June 29, 2003, B6.
22. Rowan Scarborough, "Crash Caused Lynch's 'Horrific Injuries,'" *Washington Times*, July 9, 2003.
23. John Kampfner, "The Truth About Jessica," *Guardian*, May 15, 2003.
24. Michele Orecklin, "The Controversy Over Jessica Lynch," *Time*, June 9, 2003, 33.
25. "Assessing Media Coverage," 21.
26. "Assessing Media Coverage," 23.
27. Mark Bowden, "Sometimes Heroism Is a Moving Target," *New York Times*, June 8, 2003, WR 1.
28. "Synergizing Private Lynch," *New York Times*, June 16, 2003, A12.

29. Jim Hoagland, "Clarity: The Best Weapon," *Washington Post,* June 1, 2003, B7.

30. Smith, "Hard Lessons," 28.

31. "Summary of Findings: Strong Ownership to Media Cross-Ownership Emerges," Pew Research Center for the People and the Press, July 13, 2003, people-press.org/reports/print.php3?PageID=719, 1, 2, 7.

32. Mark Jurkowitz, "Americans Want Facts and Flags," *Boston Globe,* July 14, 2003, B7.

33. Harry A. Jessell, "Protests Turn Off Viewers," *Broadcasting & Cable,* March 24, 2003, www.broadcastingcable.com/index.asp?articleID=CA286548.

34. Paul Krugman, "Behind the Great Divide," *New York Times,* February 18, 2003, A27.

35. Michael Getler, "Worth More than a One-Liner," *Washington Post,* October 6, 2002, B6.

36. Paul Farhi, "For Broadcast Media, Patriotism Pays," *Washington Post,* March 28, 2003, C1.

37. Matt Wells, "Dyke Strikes Out at U.S. Media," *Guardian,* April 25, 2003.

38. "Face Value: The First Casualty of War," *Economist,* May 3, 2003, 66.

39. "Transcript of Peter Arnett Interview on Iraqi TV," www.cnn.com/2003/WORLD/meast/03/30/sprj.irq.arnett.transcript.

40. Carol Marin, "Getting All the News Out of Baghdad," *Chicago Tribune,* April 2, 2003.

41. David Zurawik, "Peter Arnett Fired for Poor Judgment," *Baltimore Sun,* April 1, 2003.

42. Ted Gup, "Useful Secrets," *Columbia Journalism Review,* March/April 2003, 14–15.

43. Michael Getler, "Spooking, and Finding, Hussein," *Washington Post,* February 23, 2003, B6.

44. Dana Priest, "U.S. Teams Seek To Kill Iraqi Elite," *Washington Post,* March 29, 2003, A1.

45. David Zurawik, "Pentagon Losing Its Grip on News Flow," *Baltimore Sun,* March 25, 2003, 1D.

46. Eason Jordan, "The News We Kept to Ourselves," *New York Times,* April 11, 2003, A25.

47. David Folkenflik, "CNN Taking Heat for Withholding News on Iraqi Brutality," *Baltimore Sun,* April 16, 2003.

48. Bill Katovsky and Timothy Carlson, eds., *Embedded: The Media at War in Iraq* (Guilford, CT: Lyons Press, 2003), 156.

49. Peter Johnson, "CNN Takes Heat for Action, Inaction," *USA Today,* April 14, 2003, 5D.

50. Folkenflik, "CNN Taking Heat."

51. Howard Kurtz, "The Pen, Mightier Than the Minicam?" *Washington Post,* April 7, 2003, C1.

52. Rachel Smolkin, "Media Mood Swings," *American Journalism Review,* vol. 25, no. 5, June/July 2003, 18.

53. Smolkin, "Media Mood Swings," 17.

54. Smolkin, "Media Mood Swings," 23.

55. David Zurawik, "Cable Must Consider the Rest of the Story," *Baltimore Sun,* April 12, 2003, 1D.

56. Theodore L. Glasser, "The Language of War," *AEJMC News,* May 2003, 2.

57. Jason Deans, "Americans Turn to BBC for War News," *Guardian,* April 17, 2003.

Chapter Five

1. Lee Rainie, Susannah Fox, and Deborah Fallows, "The Internet and the Iraq War," Pew Internet and American Life Project, April 1, 2003, 2–3, 5, 6, www. pewinternet.org.
2. Carol Guensburg, "Online Access to the War Zone," *American Journalism Review*, vol. 21, no. 4, May 1999, 12.
3. David M. Durant, "Web Watch," *Library Journal*, September 1, 1999, 134.
4. "Screen Test," *Economist*, September 29, 2001, 64.
5. Philip Seib, ed., *New Wars, New Media* (Milwaukee: Marquette University, 2001), 100.
6. "American Web Surfers Boost Traffic to Foreign News Sites in March, According to Nielsen/NetRatings," Nielsen NetRatings, April 24, 2003, www.nielsen-netratings.com.
7. Leslie Walker, "Web Use Spikes on News of War," *Washington Post*, March 22, 2003, E1.
8. Jon Swartz, "Iraq War Could Herald a New Age of Web-based News Coverage," *USA Today*, March 19, 2003, D3.
9. Steve Outing, "War: A Defining Moment for Net News," *editorandpublisher.com*, March 26, 2003, www.editorandpublisher.com/editorandpublisher/features_columns/article_display.jsp?vnu_content_id=1848575.
10. Barb Palser, "Online Advances," *American Journalism Review*, vol. 25, no. 4, May 2003, 40.
11. Penelope Patsuris, "War May Spur Streaming News," *Forbes.com*, March 20, 2003, www.forbes.com/2003/03/20/cx_pp_0320streaming.
12. Palser, "Online Advances," 43.
13. Patsuris, "War May Spur Streaming News."
14. "Broadband Revolutionizing Europe's Internet Behavior," Nielsen NetRatings, May 29, 2003, www.nielsen-netratings.com.
15. CyberJournalist.net, "Great Iraq Conflict Coverage Gallery," www.cyberjournalist.net/features/moreiraqcoverage.
16. Howard Kurtz, "'Webloggers' Signing On as War Correspondents," *Washington Post*, March 23, 2003, F4.
17. Leslie Walker, "Operation Commentary Storm," *Washington Post*, March 23, 2003, H7.
18. Spencer E. Ante, "Have Web Site, Will Investigate," *Business Week*, July 28, 2003, 70.
19. CyberJournalist.net, "Great Iraq Conflict Coverage Gallery."
20. Rainie, Fox, and Fallows, "The Internet and the Iraq War," 5–6.
21. Mary Anne Ostrom, "Net Plays Big Role in War News, Commentary," *San Jose Mercury News*, www.siliconvalley.com/mld/siliconvalley/5285029.
22. Rory McCarthy, "Salam's Story," *Guardian*, May 30, 2003.
23. Owen Gibson, "Spin Caught in a Web Trap," *Guardian*, February 17, 2003.
24. George Packer, "Smart-Mobbing the War," *New York Times Magazine*, March 9, 2003, 46.
25. Jane Perrone, "Working the Web," *Guardian*, February 20, 2003.
26. Ken Lee, "At Last, E-Mail from Baghdad," *USA Today*, May 19, 2003.
27. Bruce Berkowitz, *The New Face of War* (New York: Free Press, 2003), 43.
28. John Arquilla and David Ronfeldt, "A New Epoch—and Spectrum—of Conflict," in John Arquilla and David Ronfeldt, eds., *In Athena's Camp: Preparing for Conflict in the Information Age* (Santa Monica: RAND, 1997), 6.

29. Robert Lemos, "What Are the Real Risks of Cyberterrorism?" *ZDNet,* August 26, 2003, zdnet.com/2100–1105–955293.

30. Barbara W. Tuchman, *The Zimmermann Telegram* (New York: Bantam, 1971), 142.

31. Ashley Dunn, "Crisis in Yugoslavia: Battle Spilling Over Onto the Internet," *Los Angeles Times,* April 3, 1999, 10.

32. Maura Conway, "Terrorism and IT: Cyberterrorism and Terrorist Organizations Online," Paper presented at the International Studies Association Annual Convention, February 27, 2003, 8–9.

33. Thom Shanker and Eric Schmitt, "Firing Leaflets and Electrons, U.S. Wages Information War," *New York Times,* February 24, 2003, A1.

34. Michelle Delio, "U.S. Tries E-mail To Charm Iraqis," *Wired,* February 13, 2003, 1.

35. Brian McWilliams, "Iraq's Crash Course in Cyberwar," *Wired,* May 22, 2003, www.wired.com/news/conflict/0,2100,58901,00.

36. Brian Krebs, "A Short History of Computer Viruses and Attacks," *washingtonpost.com,* February 14, 2003.

37. Christopher Tkaczyk, "Crushing Bugs, *Fortune,* September 15, 2003, 48.

38. Dorothy E. Denning, "Is Cyber Terror Next?," Social Science Research Council, November 1, 2001, www.ssrc.org/sept11/essays/denning, 2.

39. Dorothy E. Denning, "Activism, Hacktivism, and Cyberterrorism: The Internet as a Tool for Influencing Foreign Policy," in John Arquilla and David Ronfeldt, eds., *Networks and Netwars* (Santa Monica: RAND, 2001), 282.

40. Dickon Ross, "Electronic Pearl Harbor," *Guardian,* February 20, 2003.

41. Denning, "Activism, Hacktivism, and Cyberterrorism," 269.

42. Denning, "Is Cyber Terror Next?," 3.

43. David McGuire and Brian Krebs, "Attack on Internet Called Largest Ever," washingtonpost.com, October 22, 2002, www.washingtonpost.com/ac2/wp-dyn/A828–2002Oct22.

44. Brian Krebs and David McGuire, "More Than One Internet Attack Occurred Monday," washingtonpost.com, October 23, 2002, www.washingtonpost.com/ac2/wp-dyn/A6894–2002Oct23.

45. Barton Gellman, "Cyber-Attacks by Al Qaeda Feared," *Washington Post,* June 27, 2002, A1.

46. "Vulnerability: How Real Is the Threat?" *Frontline,* www.pbs.org/wgbh/pages/frontline/shows/cyberwar/vulnerable/threat.

47. "A Letter from Concerned Scientists," *Frontline,* www.pbs.org/wgbh/pages/frontline/shows/cyberwar/etc/letter.

48. "Vulnerability: How Real Is the Threat?"

49. Joshua Green, "The Myth of Cyberterrorism," *Washington Monthly Online,* November 2002, www.washingtonmonthly.com/features/2001/0211.green.

50. "Vulnerability: How Real Is the Threat?"

51. Lemos, "What Are the Real Risks of Cyberterrorism?"

52. Lee Rainie, "Half of Americans Fear Terrorists Might Mount Successful Cyber Attacks Against Key American Utilities and Businesses," Pew Internet Project Data Memo, August 31, 2003.

53. Michelle Delio, "Al-Qaeda Web Site Refuses To Die," *Wired,* April 7, 2003, www.wired.com/news/infostructure/0,1377,58356,00.

54. "Al-Qaeda on the Fall of Baghdad," Middle East Media Research Institute, April 11, 2003, memri.org/bin/opener.cgi?Page=archives&ID=SP49303.

55. haganah.org.il.

56. "National Strategy To Secure Cyberspace," www.whitehouse.gov/pcipb/executive_summary, viii-ix.

57. "National Strategy To Secure Cyberspace," viii, xii.

58. "Information Operations," U.S. Strategic Command Public Affairs, www.stratcom.af.mil/factsheetshtml/Information%20Operations.

Chapter Six

1. "The One Where Pooh Goes to Sweden," *Economist*, April 5, 2003, 59.

2. Jamey Keaten, "France Taking Bids for All-News Channel," *Newsday*, May 12, 2003.

3. Brian Whitaker, "Battle Station," *Guardian*, February 7, 2003.

4. James Careless, "Judging Al-Jazeera," *Communicator*, December 2002, 8.

5. Mohammed el-Nawawy and Adel Iskandar, *Al-Jazeera* (Boulder, CO: Westview, 2002), 100.

6. Neil Hickey, "Perspectives on War," *Columbia Journalism Review*, March/April 2002, 40.

7. el-Nawawy and Iskandar, *Al-Jazeera*, 20.

8. Faisal Bodi, "Al-Jazeera Tells the Truth About War," *Guardian*, March 28, 2003.

9. Jonathan Alter, "The Other Air Battle," *Newsweek*, April 7, 2003, 39.

10. Ian Black, "Television Agendas Shape Images of War," *Guardian*, March 27, 2003.

11. "Al-Jazeera Defends War Reports," BBC *Correspondent*, May 24, 2003.

12. Michael Dobbs, "In Fight for World Opinion, Results Are Mixed for U.S.," *Washington Post*, March 24, 2003, A27.

13. Dobbs, "In Fight for World Opinion," A27.

14. Hampton Sides, "A CENTCOM Star," *New Yorker*, April 21, 2003, 62.

15. S. Abdallah Schleifer, "Interview with Ibrahim Helal, Chief Editor, Al-Jazeera," *Transnational Broadcasting Studies Journal*, Spring-Summer 2003, www.tbsjournal.com/helal.

16. Dominic Timms, "Wolfowitz Sparks Fury from Al-Jazeera," *Guardian*, Tuesday, July 29, 2003.

17. E. A. Torriero, "U.S., Media at Odds Over Iraq Coverage," *Chicago Tribune*, August 1, 2003.

18. Alissa J. Rubin, "Iraqis Defend Media Ban, Allege Incitement," *Los Angeles Times*, September 25, 2003, A1.

19. el-Nawawy and Iskandar, *Al-Jazeera*, 163.

20. Peter Svensson, "Al-Jazeera Site Most Sought After," *Editor and Publisher*, April 2, 2003.

21. "All That Jazeera," *Economist*, June 21, 2003, 60.

22. "Self-doomed to Failure," *Economist*, July 6, 2002, 24, 25.

23. In'am el-Obeidi, "A Palestinian Perspective on Satellite Television Coverage of the Iraq War," *Transnational Broadcasting Studies Journal*, Spring-Summer 2003, www.tbsjournal.com/obeidi.

24. Rami G. Khouri, "The War Americans Don't See," *New York Times*, April 4, 2003, A 19.

25. Hussein Amin, "'Watching the War' in the Arab World," *Transnational Broadcasting Studies Journal*, Spring-Summer 2003, www.tbsjournal.com/amin.

26. Susan Sachs, "Arab Media Portray War as Killing Field," *New York Times*, April 4, 2003, B1.

27. "About Al-Manar Television," www.manartv.com.

28. S. Abdallah Schleifer, "Interview with Nart Y. Bouran, Director of the Abu Dhabi TV News Center," *Transnational Broadcasting Studies Journal,* Spring-Summer 2003, www.tbsjournal.com/bouran.

29. "Channel Offers Arab News," *Milwaukee Journal Sentinel,* February 20, 2003, E2.

30. "Al-Arabiya: A Balanced Alternative to Al-Jazeera?" www.arabia.com/newsfeed/print/article/english/0,14239,371610,00.

31. S. Abdallah Schleifer, "Interview with Saleh Negm, Head of News, Al-Arabiya," *Transnational Broadcasting Studies Journal,* Spring-Summer 2003, www.tbsjournal.com/negm.

32. Reuters, "Battle for Hearts of Iraqi TV Viewers Heats Up," *New York Times,* November 3, 2003, A16.

33. Stephen Franklin, "Fear of Waking Up," *Columbia Journalism Review,* September/October 2003, 60.

34. David Ignatius, "Hezbollah's Success," *Washington Post,* September 23, 2003, A27.

35. S. Abdallah Schleifer, "Interview with Ian Ritchie, CEO of Associated Press Television News," *Transnational Broadcasting Studies Journal,* Spring-Summer 2003, www.tbsjournal.com/ritchie.

36. Abdulhamid Al-Ansary, "Arab Media's Conduct During War Indicative of a Deeper Malaise," *Arab News,* April 21, 2003.

37. Amin, "'Watching the War' in the Arab World."

38. el-Obeidi, "A Palestinian Perspective."

39. Emily Wax, "Arab World Is Seeing War Far Differently," *Washington Post,* March 28, 2003, A33.

40. Middle East Media Research Institute, "Arab and Muslim Media Reactions to the Fall of Baghdad," Special Report No. 14, April 11, 2003, 4, 7.

41. Jonathan Steele and Dan De Luce, "Middle East Gets Another View of Fall of Baghdad," *Guardian,* April 10, 2003.

42. Middle East Media Research Institute, "Arab and Muslim Media Reactions," 6.

43. Middle East Media Research Institute, "Arab and Muslim Media Reactions," 1, 2.

44. Jefferson Morley, "Arab Media Confront the 'New Rules of the Game,'" *washingtonpost.com,* April 9, 2003, www.washingtonpost.com/ac2/wp-dyn/A64349–2003Apr9.

45. Dilruba Catalbas, "Divided and Confused: The Reporting of the First Two Weeks of the War in Iraq on Turkish Television Channels," *Transnational Broadcasting Studies Journal,* Spring-Summer 2003, www.tbsjournal.com/catalbas%20turkey.

46. Christine Ogan, "Big Turkish Media and the Iraq War—A Watershed?" *Transnational Broadcasting Studies Journal,* Spring-Summer 2003, www.tbsjournal.com/ogan.

47. Catalbas, "Divided and Confused."

48. www.medyatv.com.

49. Kerry Capell, "Suddenly, the BBC Is a World-beater," *Business Week,* April 28, 2003, 98.

50. Paul Reynolds, "Vox Populi—Worldwide War-Talk on the Web," in Sara Beck and Malcolm Downing, eds., *The Battle for Iraq* (London: BBC, 2003), 180, 181.

51. Robyn Dixon and Henry Chu, "Russian, European Media Critical of U.S.-Led Forces," *Los Angeles Times,* April 3, 2003.

52. Mark Landler, "German News Channels Come of Age," *New York Times,* March 31, 2003.

53. Jefferson Morley, "The Rest of the West Is Less than Impressed," *washington-post.com*, April 10, 2003.

54. Morley, "The Rest of the West."

55. "Opinions on the Iraq War from Newspapers Around the World," *New York Times*, April 6, 2003, B12.

56. Dixon and Chu, "Russian, European Media Critical."

57. "Opinions on the Iraq War."

58. Ellen Nakashima, "In Indonesia, a Wary Worldview," *Washington Post*, April 8, 2003, C1.

59. Richard Lambert, "Misunderstanding Each Other," *Foreign Affairs*, March/April 2003, 72.

60. Lambert, "Misunderstanding Each Other," 67–70.

61. "Opinions on the Iraq War."

62. Anthony Borden, "Help Them Create a BBC of Their Own," *Washington Post National Weekly Edition*, April 21, 2003, 22.

63. Neela Banerjee, "Iraqis Race To Fill Void in Journalism," *New York Times*, May 13, 2003.

64. Banerjee, "Iraqis Race."

65. Sabrina Tavernise, "Iranian News Channel Makes Inroads in Iraq," *New York Times*, April 29, 2003.

66. "A New Voice in the Middle East: A Provisional Needs Assessment for the Iraqi Media," report prepared by the Baltic Media Centre, Index on Censorship, Institute for War and Peace Reporting, and International Media Support, www.iwpr.net/pdf/Iraq_Media_Assessment_Report.pdf, 5–10.

67. Torriero, "U.S., Media at Odds."

68. "A New Voice in the Middle East," 7.

Chapter Seven

1. Elizabeth Becker, "The American Portrayal of a War of Liberation Is Faltering Across the Arab World," *New York Times*, April 5, 2003, B10.

2. Michael Dobbs and Mike Allen, "Images of Destruction Inflict Setback for U.S. Propaganda War," *Washington Post*, March 30, 2003, A26.

3. Christopher Ross, "Pillars of Public Diplomacy," *Harvard International Review*, vol. xxv, no. 2, Summer 2003, 22.

4. Stephen Johnson and Helle Dale, "How To Reinvigorate U.S. Public Diplomacy," *Heritage Foundation Backgrounder*, April 23, 2003, www.heritage.org/research/nationalsecurity/bg1645.

5. Stephen Hess and Marvin Kalb, eds., *The Media and the War on Terrorism* (Washington: Brookings Institution, 2003), 225.

6. Jim Hoagland, "The MIA State Department," *Washington Post*, March 30, 2003, B7.

7. Andrew Kohut, "American Public Diplomacy in the Islamic World," remarks to the Senate Foreign Relations Committee hearing, February 27, 2003, people-press.org/commentary/print.php?AnalysisID=63.

8. Shibley Telhami, "The Need for Public Diplomacy," *Brookings Review*, Summer 2002, 48.

9. Karen DeYoung, "Bush To Create Formal Office To Shape U.S. Image Abroad," *Washington Post*, July 30, 2002, A1.

10. "Finding America's Voice: A Strategy for Reinvigorating U.S. Public Diplomacy," Report of an independent task force sponsored by the Council on Foreign Relations, September 2003, www.cfr.org/pdf/public_diplomacy.pdf., 3.

11. "Finding America's Voice," 16.

12. "Transatlantic Trends 2003 Survey Results Released," German Marshall Fund of the United States news release, September 4, 2003, www.gmfus.org/apps/gmf/gmfwebfinal.nsf.

13. Karen DeYoung, "Bush Message Machine Is Set To Roll with Its own War Plan," *Washington Post,* March 19, 2003, A 1.

14. Bob Kemper, "Agency Wages Media Battle," *Chicago Tribune,* April 7, 2003.

15. Michael Dobbs, "Envoy to 'Arab Street' Stays Hopeful," *Washington Post,* June 10, 2003, A19.

16. Charlotte Beers, "American Public Diplomacy and Islam," testimony before the Senate Foreign Relations Committee, February 27, 2003, www.state.gov/r/us/18098.

17. Terence Smith, "Public Diplomacy," *NewsHour,* Public Broadcasting System, January 21, 2003.

18. Johnson and Dale, "How To Reinvigorate U.S. Public Diplomacy," 2; "Finding America's Voice," 8–16.

19. Michael White, "Blair and Bush Broadcast TV Messages to Iraqis," *Guardian,* April 11, 2003.

20. Fawn Vrazo, "Freedom TV Tries To Define Itself," *Philadelphia Inquirer,* April 14, 2003.

21. Ciar Byrne, "Blair Launches New Iraqi TV," *Guardian,* April 10, 2003.

22. Karen DeYoung and Walter Pincus, "U.S. To Take Its Message to Iraqi Airwaves," *Washington Post,* May 11, 2003, A17.

23. Smith, "Public Diplomacy."

24. Anthony Shadid, "Seeking a Voice in the Arab World, U.S. Tries Radio," *Boston Globe,* August 7, 2002.

25. Shane Harris, "Government Will Launch News Network To Counter 'Anti-American' Image," *Government Executive,* June 2, 2003, www.govexec.com/news/index.cfm?mode=report&article=25753.

26. Lynette Clemetson with Nazila Fathi, "U.S.'s Powerful Weapon in Iran: TV," *New York Times,* December 7, 2002, A21.

27. Nazila Fathi, "TV Stations Based in U.S. Rally Protestors in Iran," *New York Times,* June 22, 2003, A8.

28. www.voa.gov/index.

29. Karen DeYoung, "France Says It Is Target of Untruths," *Washington Post,* May 15, 2003, A1.

30. "New Strategic Direction Urged for Public Diplomacy," Advisory Commission on Public Diplomacy press release, U.S. Department of State, October 1, 2003, www.state.gov/r/adcompd/rls/prls/24777.

31. Anne E. Kornblut, "Problems of Image and Diplomacy Beset U.S.," *Boston Globe,* March 9, 2003, A25.

32. David Ignatius, "America's Doubters in Beirut," *Washington Post,* June 6, 2003, A27.

33. Nicholas D. Kristof, "Saudis in Bikinis," *New York Times,* October 25, 2002.

34. Joseph Nye, *The Paradox of American Power,* (New York: Oxford, 2002), 9, 141.

35. Nye, *The Paradox of American Power,* 11.

Chapter Eight

1. John Barry, "War Costs," *Newsweek,* May 26, 2003, 10.
2. Nancy Gibbs with Mark Thompson, "A Soldier's Life," *Time,* July 21, 2003, 29–33.
3. George W. Bush, Remarks at the 20th Anniversary of the National Endowment for Democracy, United States Chamber of Commerce, Washington, D.C., November 6, 2003, www.whitehouse.gov/news/releases/2003/11/print/20031106–2, 4.
4. Seth Mnookin, "War Reports," *Newsweek,* www.msnbc.com/news/903675.
5. "Assessing Media Coverage of the War in Iraq," 28.
6. Peter J. Boyer, "The New War Machine," *New Yorker,* June 30, 2003, 60, 64.
7. Evan Thomas and Martha Brant, "The Education of Tommy Franks," *Newsweek,* May 19, 2003, 27.
8. Boyer, "The New War Machine," 69.
9. Michael Duffy, Mark Thompson, and Michael Weisskopf, "Secret Armies of the Night," *Time,* June 23, 2003, 42.
10. Duffy, Thompson, and Weisskopf, "Secret Armies," 45.
11. Boyer, "The New War Machine," 61, 70.
12. "The Janus-faced War," *Economist,* April 26, 2003, 24.
13. Duffy, Thompson, and Weisskopf, "Secret Armies," 45.
14. "The Janus-faced War," 24.
15. David Ignatius, "Standoffish Soldiering," *Washington Post,* August 5, 2003, A15.
16. Wesley K. Clark, *Winning Modern Wars* (New York: PublicAffairs, 2003), 92.
17. Robert D. Kaplan, "Supremacy by Stealth," *Atlantic Monthly,* July/August 2003, 66.
18. Kaplan, "Supremacy by Stealth," 76.
19. George Robertson, "Time To Deliver," *World Today,* vol. 59, no. 6, June 2003, 12.
20. Robertson, "Time To Deliver," 11.
21. Vaclav Havel, "An Alliance with a Future," *Washington Post,* May 19, 2002, B7.
22. George W. Bush, Remarks at the Royal Banqueting House, Whitehall Palace, London, November 19, 2003, www.whitehouse.gov/news/releases/2003/11/print/20031119–1, 2.
23. Tony Blair, "Doctrine of International Community," speech delivered to the Economic Club of Chicago, April 22, 1999, www.globalpolicy.org/globaliz/politics/blair.
24. Ivo H. Daalder and James M. Lindsay, "The Globalization of Politics," *Brookings Review,* Winter 2003, 14.
25. Thomas L. Friedman, *The Lexus and the Olive Tree* (New York: Farrar Straus Giroux, 1999), 7.
26. Kurt M. Campbell, "Globalization's First War?" *Washington Quarterly,* Winter 2002, 8.
27. Bruce Hoffman, "The Logic of Suicide Terrorism," *Atlantic Monthly,* June 2003, 40.
28. C. A. J. Coady, "The Ethics of Armed Humanitarian Intervention," United States Institute of Peace, *Peaceworks* 45, July 2002, 6.
29. George W. Bush, Remarks at the Royal Banqueting House, Whitehall Palace, London, November 19, 2003, 3.
30. Coady, "The Ethics of Armed Humanitarian Intervention," 35–36.
31. David Rieff, "Were Sanctions Right?" *New York Times Magazine,* July 27, 2003, 43.

32. "The Global Menace of Local Strife," *Economist*, May 24, 2003, 23.
33. "The Global Menace of Local Strife," 24.
34. "The Poor Man's Curse," *Economist*, May 24, 2003, 11.
35. David Ignatius, "Turning Africa Around," *Washington Post*, June 10, 2003, A21.
36. Ignatius, "Turning Africa Around."
37. "Peace, They Say, But the Killing Goes On," *Economist*, March 29, 2003, 41.
38. Philip Gourevitch, "Forsaken," *New Yorker*, September 25, 2000, 65.
39. Emily Wax, "Boy Soldiers Toting AK–47s Put at Front of Congo's War," *Washington Post*, June 14, 2003, A1.
40. Thomas Fuller, "European Peacekeepers to Go to Congo on Non-NATO Mission," *New York Times*, June 5, 2003, A5.
41. Pierre Conesa, "A Pedigree for the Dogs of War," *Le Monde Diplomatique*, April 2003, 7.
42. Kofi Annan, *The Question of Intervention* (New York: United Nations Department of Public Information, 1999), 13.
43. David Chandler, *From Kosovo to Kabul: Human Rights and International Intervention* (London: Pluto Press, 2002), 58, 60.
44. Michael Ignatieff, *The Warrior's Honor: Ethnic War and the Modern Conscience* (New York: Henry Holt, 1997), 10, 11, 23.
45. Howard Kurtz, "For Media After Iraq, A Case of Shell Shock," *Washington Post*, April 28, 2003, A1.
46. Kurtz, "For Media After Iraq," A1; Lori Robertson, "Whatever Happened to Afghanistan?" *American Journalism Review*, vol. 25, no. 5, June/July 2003, 25.
47. Robertson, "Whatever Happened to Afghanistan?" 29.
48. Robertson, "Whatever Happened to Afghanistan?" 28.
49. Michael Getler, "Supporting Troops, Assessing Coverage," *Washington Post*, May 4, 2003, B6.
50. Michael Getler, "The Next Casualty?" *Washington Post*, June 15, 2003, B6.

SELECTED BIBLIOGRAPHY

Books

Anderson, Jon Lee. *The Lion's Grave: Dispatches from Afghanistan*. New York: Grove, 2002.

Arquilla, John and David Ronfeldt, eds. *In Athena's Camp: Preparing for Conflict in the Information Age*. Santa Monica: RAND, 1997.

Aukofer, Frank and William P. Lawrence. *America's Team: The Odd Couple*. Nashville: Freedom Forum First Amendment Center, 1995.

Beck, Sara and Malcolm Downing, eds. *The Battle for Iraq*. London: BBC, 2003.

Berkowitz, Bruce. *The New Face of War*. New York: Free Press, 2003.

Bobbitt, Philip. *The Shield of Achilles: War, Peace, and the Course of History*. New York: Knopf, 2002.

Bunt, Gary R. *Islam in the Digital Age*. London: Pluto Press, 2003.

Burke, Jason. *Al-Qaeda*. London: I. B. Tauris, 2003.

Dobkin, Bethami. *Tales of Terror: Television News and the Construction of the Terrorist Threat*. Westport, CT: Praeger, 1992.

el-Nawawy, Mohammed and Adel Iskander. *Al-Jazeera*. Boulder, CO: Westview, 2002.

Esposito, John L. *Unholy War: Terror in the Name of Islam*. New York: Oxford, 2002.

Glennon, Michael. *Limits of Law, Prerogatives of Power: Intervention After Kosovo*. New York: Palgrave Macmillan, 2001.

Hedges, Chris. *War Is a Force that Gives Us Meaning*. New York: PublicAffairs, 2002.

Hess, Stephen and Marvin Kalb, eds. *The Media and the War on Terrorism*. Washington: Brookings Institution, 2003.

Hoffman, Bruce. *Inside Terrorism*. New York: Columbia University Press, 1998.

Huntington, Samuel P. *The Clash of Civilizations and the Remaking of World Order*. New York: Simon and Schuster, 1996.

Ignatieff, Michael. *The Warrior's Honor: Ethnic War and the Modern Conscience*. New York: Henry Holt, 1997.

Kagan, Robert. *Of Paradise and Power*. New York: Knopf, 2003.

Kaldor, Mary. *New and Old Wars: Organized Violence in a Global Era*. Cambridge, UK: Polity Press, 1999.

Kaplan, Robert D. *The Coming Anarchy*. New York: Random House, 2000.

———. *The Ends of the Earth*. New York: Vintage, 1997.

Katovsky, Bill and Timothy Carlson, eds. *Embedded: The Media at War in Iraq*. Guilford, CT: Lyons Press, 2003.

Kupchan, Charles. *The End of the American Order*. New York: Knopf, 2002.

Laqueur, Walter. *No End to War*. New York: Continuum, 2003.

MacMillan, Margaret. *Paris 1919*. New York: Random House, 2002.

Moeller, Susan D. *Compassion Fatigue*. New York: Routledge, 1999.

Nacos, Brigitte. *Mass-mediated Terrorism*. Lanham, MD: Rowman and Littlefield, 2002.

Nye, Joseph. *The Paradox of American Power*. New York: Oxford, 2002.

O'Meara, Patrick, Howard D. Mehlinger, and Matthew Krain, eds. *Globalization and the Challenges of a New Century*. Bloomington, IN: Indiana University Press, 2000.

Pollack, Kenneth M. *The Threatening Storm: The Case for Invading Iraq*. New York: Random House, 2002.

Sebald, W. G. *On the Natural History of Destruction*. New York: Random House, 2003.

Seib, Philip. *The Global Journalist: News and Conscience in a World of Conflict*. Lanham, MD: Rowman and Littlefield, 2002.

————, ed. *New Wars, New Media: Covering Armed Conflict Since the Gulf War*. Milwaukee: Marquette University, 2001.

Sifry, Micah L. and Christopher Cerf. *The Iraq War Reader*. New York: Touchstone, 2003.

Singer, Peter. *One World*. New Haven: Yale University Press, 2002.

Sontag, Susan. *Regarding the Pain of Others*. New York: Farrar, Straus and Giroux, 2003.

Woolf, Virginia. *Three Guineas*. New York: Harcourt, 1938.

Articles, Reports, Et Cetera

Alter, Jonathan. "The Other Air Battle," *Newsweek*, April 7, 2003, 39.

"America and the World: The Impact of September 11 on U.S. Coverage of International News," survey by Dwight L. Morris and Associates, conducted for the Pew International Journalism Program, June 2002.

Arnett, Peter. "Transcript of Interview on Iraqi TV." www.cnn.com/2003/WORLD/meast/03/30/sprj.irq.arnett.transcript.

"Assessing Media Coverage of the War in Iraq." Brookings Institution forum, June 17, 2003, www.brookings.edu/comm/events/20030617.pdf.

Beers, Charlotte. "American Public Diplomacy and Islam." Testimony before the Senate Foreign Relations Committee, February 27, 2003, www.state.gov/r/us/18098.

Bowden, Mark. "Sometimes Heroism Is a Moving Target," *New York Times*, June 8, 2003, WR1.

Boyer, Peter. "The New War Machine" *New Yorker*, June 30, 2003, 55–71.

Capell, Kerry. "Suddenly, the BBC Is a World-Beater," *Business Week*, April 28, 2003, 98.

Coady, C. A. J., "The Ethics of Armed Humanitarian Intervention," United States Institute of Peace, *Peaceworks* 45, July 2002.

Denning, Dorothy. "Is Cyber Terror Next?," Social Science Research Council, November 1, 2001, www.ssrc.org/sept11/essays/denning.

DeParle, Jason. "Long Series of Military Decisions Led to Gulf War News Censorship," *New York Times*, May 5, 1991, A1.

Dobbs, Michael. "In Fight for World Opinion, Results Are Mixed for U.S.," *Washington Post*, March 24, 2003, A27.

Duffy, Michael, Mark Thompson, and Michael Weisskopf. "Secret Armies of the Night," *Time*, June 23, 2003, 40–45.

Farhi, Paul. "For Broadcast Media, Patriotism Pays," *Washington Post*, March 28, 2003, C1.

Fleeson, Lucinda. "Bureau of Missing Bureaus," *American Journalism Review*, vol. 25, no. 7, October/November 2003, 32–39.

Gellman, Barton. "Cyber-Attacks by Al-Qaeda Feared," *Washington Post*, June 27, 2002, A1.

Getler, Michael. "A Long, Incomplete Correction," *Washington Post*, June 29, 2003, B6.

————. "Reporting Private Lynch," *Washington Post*, April 20, 2003, B6.

————. "Supporting Troops, Assessing Coverage," *Washington Post,* May 4, 2003, B6.

Glasser, Theodore L. "The Language of War," *AEJMC News,* May 2003, 2.

"Great Iraq Conflict Coverage Gallery." www.cyberjournalist.net/features/moreiraqcoverage.

Green, Joshua. "The Myth of Cyberterrorism," *Washington Monthly Online,* November 2002, www.washingtonmonthly.com/features/2001/0211.green.

Grossman, Lawrence K. "War and the Balance Sheet," *Columbia Journalism Review,* May/June 2003, 6.

Ignatieff, Michael. "The Burden," *New York Times Magazine,* January 5, 2003, 22–28.

Ignatius, David. "Turning Africa Around," *Washington Post,* June 10, 2003, A21.

Johnson, Steve. "Realities of War Put News Ethics to Test," *Chicago Tribune,* March 26, 2003, C3.

Jordan, Eason. "The News We Kept to Ourselves," *New York Times,* April 11, 2003, A25.

Jurkowitz, Mark. "Americans Want Facts and Flags," *Boston Globe,* July 14, 2003, B7.

Kampfner, John. "The Truth About Jessica," *Guardian,* May 15, 2003.

Kaplan, Robert. "Supremacy by Stealth," *Atlantic Monthly,* July/August 2003, 66–83.

Kennicott, Philip. "The Illustrated Horror of Conflict," *Washington Post,* March 25, 2003, C1.

Kohut, Andrew. "American Public Diplomacy in the Islamic World." Remarks to the Senate Foreign Relations Committee hearing, February 27, 2003, people-press.org/commentary/print.php?AnalysisID=63.

Kurtz, Howard. "For Media After Iraq, A Case of Shell Shock," *Washington Post,* April 28, 2003, A1.

Lambert, Richard. "Misunderstanding Each Other," *Foreign Affairs,* March/April 2003, 62–74.

Lemann, Nicholas. "The Next World Order," *New Yorker,* April 1, 2002, 42–48.

Lewis, Justin. "Facts in the Line of Fire," *Guardian,* November 6, 2003.

MacArthur, John. "The Lies We Bought," *Columbia Journalism Review,* May/June 2003, 62–63.

McCarthy, Rory. "Salam's Story," *Guardian,* May 30, 2003.

Middle East Media Research Institute, "Arab and Muslim Media Reactions to the Fall of Baghdad." Special Report No.14, April 11, 2003, www.memri.org/bin/articles.cgi?Page=archives&Area=sr&ID=sr01403.

"National Strategy To Secure Cyberspace." www.whitehouse.gov/pcipb.

Nelson, Scott Bernard. "Embedded Reporter Comes Away from Front Lines Torn," *Boston Globe,* April 22, 2003, E1.

Packer, George. "Smart-mobbing the War," *New York Times Magazine,* March 9, 2003, 46–49.

Palser, Barb. "Online Advances," *American Journalism Review,* May 2003, 40.

Project for Excellence in Journalism, "Embedded Reporters: What Are Americans Getting?" April 3, 2003, www.journalism.org/resources/research/reports/war/embed.

————, "The War on Terrorism: The Not So New Television News Landscape," May 23, 2002, www.journalism.org/resources/research/reports/landscape.

"Public Affairs Guidance on Embedding Media During Possible Future Operations/Deployments in the U.S. Central Command's Area of Responsibility," United States Department of Defense, www.defenselink.mil/news/Feb2003/d20030228pag.

Rainie, Lee, Susannah Fox, and Deborah Fallows. "The Internet and the Iraq War," Pew Internet and American Life Project, April 1, 2003, www.pewinternet.org/reports/toc.asp?Report=87.

Ricchiardi, Sherry. "Close to the Action," *American Journalism Review*, May 2003, 28–35.

Robertson, George. "Time To Deliver," *World Today*, June 2003, 11–12.

Robertson, Lori. "Whatever Happened to Afghanistan?," *American Journalism Review*, June/July 2003, 24–31.

Ross, Christopher. "Pillars of Public Diplomacy," *Harvard International Review*, Summer 2003, 22–27.

Rutenberg, Jim. "Ex-Generals Defend Their Blunt Comments," *New York Times*, April 2, 2003, B9.

Rutenberg, Jim and Bill Carter. "Spectacular Success or Incomplete Picture?," *New York Times*, April 20, 2003, A1.

Schmidt, Susan and Vernon Loeb. "'She Was Fighting to the Death': Details Emerging of W. Va. Soldier's Capture and Rescue," *Washington Post*, April 3, 2003, A1.

Sharkey, Jacqueline. "The Television War," *American Journalism Review*, May 2003, 18–27.

Simon, Steven. "The New Terrorism," *Brookings Review*, Winter 2003, 18–24.

Smith, Terence. "Hard Lessons," *Columbia Journalism Review*, May/June 2003, 26–28.

Smolkin, Rachel. "Media Mood Swings," *American Journalism Review*, June/July 2003, 16–23.

Swartz, Jon. "Iraq War Could Herald a New Age of Web-based News Coverage," *USA Today*, March 19, 2003, D3.

Zurawik, David. "Barrage of Military Opinions," *Baltimore Sun*, March 31, 2003.

INDEX